FOOD
FIGHT

FOOD FIGHT

GMOS AND THE FUTURE OF THE AMERICAN DIET

McKay Jenkins

AVERY

an imprint of Penguin Random House

NEW YORK

AVERY

An imprint of Penguin Random House LLC
375 Hudson Street
New York, New York 10014

Most Avery books are available at special quantity discounts for bulk purchase
for sales promotions, premiums, fund-raising, and educational needs.
Special books or book excerpts also can be created to fit specific needs.
For details, write SpecialMarkets@penguinrandomhouse.com.

Library of Congress Cataloging-in-Publication Data

Names: Jenkins, McKay, 1963- author.
Title: Food fight : GMOs and the future of the American diet / McKay Jenkins.
Description: New York : Avery, 2017. | Includes bibliographical references
and index.
Identifiers: LCCN 2016054194 (print) | LCCN 2016056950 (ebook) | ISBN
9781594634604 (hardcover) | ISBN 9780698409835 (epub)
Subjects: LCSH: Transgenic plants. | Crops—Genetic engineering.
Classification: LCC SB123.57 .J46 2017 (print) | LCC SB123.57 (ebook) | DDC
631.5/233—dc23
LC record available at https://lccn.loc.gov/2016054194

ISBN 9781101982204 (paperback)

Printed in the United States of America
1   3   5   7   9   10   8   6   4   2

Book design by Jane Treuhaft

For my teachers, my students,
and my family

CONTENTS

Part Three
FRUIT

FOOD
FIGHT

Square Tomatoes

Back in 1994, when I was pulling down four bucks an hour grading papers and teaching college students how to write, a friend told me about a can't-lose investment scheme that was sure to lift me from my economic doldrums.

Forget about investing in Amazon.com, he said. Here's what you need to get into: Square tomatoes.

They're going to be great, he said breathlessly. They've had their genes altered by scientists! They stay ripe longer, and soften more slowly, and because they're square, they can be stacked for shipping, which will bring transportation costs way down. It's like the laboratory has taken nature and made it better!

The company that makes them will make a fortune, my friend said. And so will we!

There was much truth to what my friend told me, and a good bit of misinformation as well. The product in question turned out to be the Flavr Savr tomato, a newfangled plant designed by a biotech company called Calgene. The Flavr Savr had indeed been designed not for exquisite taste, or enhanced nutrition, but to plug into an industrial food system already rapidly replacing traditional farming practices. Forget small farmers selling their fruit to their neighbors; this was big business. That year, 4 billion dollars' worth of industrial tomatoes were being picked (and shipped) while still hard and green, then reddened with ethylene gas before hitting the supermarket shelves like crates of billiard balls. The genetically altered Flavr Savr, by contrast, was designed to ripen on the vine, but was still tough enough to resist rotting. This meant it could survive both mechanical harvesting and the thousand-mile truck to market.

In 1994, after three years of negotiations with government regulators, the Flavr Savr became the first genetically modified food approved by the U.S. Food and Drug Administration (FDA) to be sold in the American supermarkets. Rather than being declared formally "safe," the Flavr Savr was considered the "substantial equivalent" of a normal tomato. At the time, few people complained, and suspicious critics of genetic engineering were largely drowned out by cheerleaders in industry and the press. Connie Chung, Jane Pauley, and Katie Couric all reported on the Flavr Savr on national television; on *NBC Nightly News*, Tom Brokaw said the tomato "stays riper, longer than the nonengineered variety, and they say it's tastier."

To be honest, as a budding English professor, I could never muster much enthusiasm for a product spelled Flavr Savr. The phonetically engineered name offended my ear even before I considered the tomato's provenance or taste, or the many ethical questions surrounding its creation. I decided to save my money, and keep grading papers.

But the Flavr Savr, it turned out, was just the beginning of what would become a food revolution. Soon I started hearing stories about another tomato, this one created by a company called DNA Plant Technology, which was being outfitted with genes from an

Arctic flounder. These "fish tomatoes," the company hoped, would make plants resistant to frost and cold storage, making them easier to grow in northern climates.

In 2001, researchers at the University of California, Davis, and the University of Toronto unveiled a third tomato, this one capable of growing in salty soils—a good thing, since modern irrigation practices were damaging soil so much that the world was losing 25 million acres of cropland a year.

The fish tomatoes never made it to market. So far, neither have the salt-tolerant tomatoes. The Flavr Savr tomatoes made it to market briefly, but they were a commercial flop; the agrochemical giant Monsanto bought the company in 1996, and dropped the product. The ingenuity of a human-engineered tomato never quite overcame the consensus that the Flavr Savrs tasted terrible. As for the Flavr Savr being square? Well, that turned out to be untrue. Blocky tomatoes had in fact been cultivated by California plant breeders in the 1950s, to make mechanical harvesting easier and to prevent them from rolling off conveyor belts, but squareness was never part of the Flavr Savr profile. This myth was just the first of what would become a long series of myths that continue to tangle themselves around engineered food like aggressive vines.

Now, more than twenty years later, these moribund tomato experiments seem almost quaint. Today, nearly all of our calories— that is to say, nearly all of our food—are grown from genetically modified plants. Chances are that three-quarters of everything you've put in your mouth today—the eggs, the yogurt, and the cereal; the chicken sandwich, the tortilla chips, the mayonnaise, and the salad dressing; the cheeseburger, the french fries, the soda, the cookies, and the ice cream—were processed (or fed) from plants grown from seeds engineered in a laboratory. Same for the food you feed your baby and the food you feed your dog.

The reason for this is simple: The American diet is composed almost entirely of processed foods that are made from two plants— corn and soybeans (and canola, if you want your food fried). Their seeds, full of dense calories, can be broken down and reconstituted into an infinite variety of prepackaged foods. The vast majority of the 40,000 food products Americans choose from every day are built from ingredients made from engineered plants. This includes almost anything made with high-fructose corn syrup, vegetable oil, or sugar—which is to say, almost all processed food. They can also be ground up and fed to the animals who provide our boundless appetite for meat and dairy products. Fully 85 percent of the feed given to cattle, hogs, and chickens is grown from genetically modified crops. There's more: About half of the sugar we consume is grown from engineered sugar beets. Genetically modified wheat has not yet hit the commercial market, but some of the biggest seed and chemical companies in the world have been working on it for years and have it ready to go.

Strangely—and despite the fact that we're talking about plants— the one place you mostly *won't* find engineered food is in the produce aisle. Your carrots, your peaches, your lettuce—they are all grown the old-fashioned way. (This, by the way, is true whether or not the produce is labeled "organic.") But travel to the middle of your supermarket—or into most fast-food restaurants, convenience stores, or gas stations—and you will discover GM foods at every turn.

Depending on whom you ask, "genetically modified organisms," or more simply "GMOs," represent either a great stride forward in the history of food production or are part of a destructive and dangerous system that allows global food companies to radically damage our land and water, control the way we eat, and flood our bodies with unhealthy food.

At the most basic level, genetic engineering is a crop-improvement technique, one of many used by plant growers, to alter the quantity, quality, and usefulness of the plants used to make food. A GMO is a plant grown from a seed genetically engineered to express a specific set of traits. These traits can range from an increased tolerance to floods or drought (a critical need given rising global temperatures) to beneficial nutrients (like rice that produces its own beta-carotene) to an improved resistance to certain viruses or insects. Such experiments—often designed by scientists at universities or nonprofit research centers—hold tremendous potential for improving the lives of people around the world. Childhood blindness in Asia, insect infestations in Africa, famines caused by typhoons in the Indian subcontinent: all are problems being addressed by GMO researchers around the world.

But it is also true that the giant agrochemical companies that produce the vast majority of the world's GMOs do very little of this work—despite their frequent claims that GMO technology can feed the world. These companies, like their cousins in the pharmaceutical industry, are far more interested in creating billion-dollar products for the American consumer market than they are in developing products—cassava, rice, sorghum—that people in the developing world actually eat. In fact, just one-half of 1 percent of American food exports actually goes to developing countries with dire food needs, a recent study by the Environmental Working Group shows. Fully 86 percent goes to wealthy, highly developed countries in Europe, Canada, Australia, Japan, and South Korea. Indeed, far from solving problems, GMO-based industrial farming actually *contributes* both to a wide variety of health problems, like obesity, diabetes, nutritional deficiency, and exposure to pesticides, and ecological problems, like water pollution, soil depletion, and a profound drop in the biodiversity of plants, animals, and insects. There's a reason the monarch butterfly has become a symbol for anti-GMO activists: Monarch food supplies have been erased by chemical sprays applied to hundreds of millions of acres of mono-culture GM crops. Nationwide, monarch populations are down by 96 percent. So when companies say GMOs are necessary to "feed a starving world," the slogan can sound empty, cynical, a bait and switch.

In the United States, and increasingly in the developing world, GMOs are planted not to improve global nutrition but to maximize corporate profits through the production of corn and soybeans, which are then funneled into a global system of processed food and industrial meat. In order to support production, they are engineered to tolerate vast quantities of chemical sprays, which are often made by the same companies that make the seeds themselves. These sprays significantly damage both human health and environmental integrity. And because only large companies can fund most GMO research and development, they patent any seeds they create, which means they can control how and by whom they are used. Since time immemorial, farmers developed, saved, and traded seeds from one year to the next, bartering their way to better, more fruitful crops. No more. Now, GM seed companies force farmers to sign agreements that they will not save or share seeds, and hire investigators to badger (or sue) them when they do. As a result, our food supply is essentially controlled by a very small number of enormous biotech companies, most of which got their start making explosives, plastics, and pesticides.

This trend has given rise to a symbiotic but imbalanced relationship between these companies and our government. Because of their size and power, companies hold tremendous sway over federal food policy, from the way food and chemicals are (or are not) regulated to what kinds of farms (and food companies) receive hundreds of billions of dollars in federal subsidies to how much information companies need to disclose about their processes and products. The companies that design and sell GM seeds are some of the biggest in the world, and yet they are oddly invisible. You may have heard the names Monsanto, DuPont, and Dow, but these company names do not appear anywhere on your cereal box. In 2009, the top six agro-chemical companies (Monsanto, DuPont, and Dow, plus Syngenta, Bayer, and BASF) earned a combined $27.4 billion in seed sales and $44.4 billion in chemical sales. Collectively, they control two-thirds of the world's agrochemical market. By 2019, the global agrochemi-cal industry is expected to reach a value of $261 billion. And since several of the biggest companies are in the process of merging, their influence will soon be consolidated further.

The closer you look at the GMO debate, the more you are con-fronted with questions and paradoxes and passionate believers on all sides. Take, for example, the seemingly innocuous question "Are GMOs safe?" A great many scientists say altering a plant's genes in a laboratory is merely one incremental improvement in a long his-tory of plant breeding, that GMOs are among the most studied—and thus the safest—foods ever produced, and that there is absolutely nothing to worry about. A library of scientific reports, and reputa-ble organizations like the National Academy of Sciences, support this claim.

But such pronouncements are less than entirely satisfying, given that many GM crops are grown (indeed, are designed) to be sprayed with hundreds of millions of pounds of petrochemical insecticides (to kill bugs) and herbicides (to kill weeds). Whether or not geneti-cally altered seeds themselves are benign, the chemicals that accom-pany them are not. The World Health Organization recently declared glyphosate, an herbicide sprayed on Monsanto's Roundup

Ready food crops around the world and long considered a relatively tame herbicide, to be a "probable human carcinogen."

The agrochemical companies—and the giant food-processing companies they supply with grains—have argued vigorously that GMOs are entirely safe. Yet when market demands change—when consumers express fears about GMOs—some of these same companies boast long and loud when they remove them from their products. In a nod to anxious mothers, the Hershey Company says it will stop using GM sugar beets to make its milk chocolate and Hershey's Kisses. General Mills will stop using GM ingredients in Cheerios and recently announced it will label any of its products that contain GMOs. Del Monte, one of the country's biggest producers of canned fruits and vegetables, says it will cease using GM ingredients in most of its products.

McDonald's refuses to sell GM potatoes grown by J. R. Simplot, one of its biggest french fry suppliers. Do these moves constitute a stance on GMOs, or only a desire to satisfy a nervous market? Hard to say. The meat McDonald's sells is still raised on GM corn, and the soda it sells is still sweetened with GM corn syrup. Cheerios are made mostly of oats, which are never grown with GMOs, so the only change General Mills really has to make is to replace sweeteners made from GM sugar beets and cornstarch made from GM corn. And the company has made it plain that it will continue using GMOs in its other cereals.

A lot of consumers who pay close attention to the GMO debate are convinced that GMOs are in fact unsafe, and a great many of them shop in stores that take advantage of this anxiety. Whole Foods, Trader Joe's, Chipotle—these national chains have all made a fuss about going "GMO free" to one degree or another. Are these claims trustworthy, or are they merely marketing schemes?

In Europe, centuries of intertwined, small, local farms have made GMOs a thing of almost continental contempt. As hard as they have tried, giant food conglomerates have had a tough time persuading the French, and the Italians, and the Spanish, to give over their land—and their diets—to industrial corn and soybeans. Globally, there are currently twenty-six countries with total or partial bans on GMOs, including Australia, China, India, Mexico, and Russia. In early 2015, thousands of Polish farmers drove their tractors into the streets in Warsaw to push for a ban on GMOs and to fight a perceived land grab by big ag-biotech companies like Monsanto. "The health and welfare of the nation depends on consumers and farmers having access to traditional seeds and good-quality food," one farmer said. "The Polish government does not accept this and is destroying the roots of Polish agriculture by listening to corporations rather than the Polish people."

For a variety of reasons, such heat has not been present in the United States. Drive across the country, as my family and I did last summer, and you will find yourself crossing a continent almost entirely given over to corn and soybeans. Granted, there are boundless waves of wheat growing across the northern Midwest, but the route we took—from the Eastern Shore of Maryland to the Colorado Rockies, then up to the Olympic Peninsula in Washington state— was astonishingly dichromatic. For close to 4,500 miles, my wife and I would switch off driving and snoozing, our kids in the backseat listening to audio books. This was hardly scientific, but our experience was absolutely clear: you can fall asleep passing fields of corn in Ohio and wake up passing fields of soy in Indiana, or vice versa, but that's about it. Only once during the whole cross-country trip did we find ourselves surprised by what we saw: a large farm in Virginia that was actually growing potatoes.

Despite all the romantic rhetoric thrown around about farming in America, it's hard to feel sentimental when all the land you see, for thousands and thousands of miles, is being used to grow corn and soy for cheap chicken and cattle feed, or frying oil, or salty snacks, or ethanol for gasoline. These crops—hundreds of millions of acres of them, and virtually all GMO—are grown far from population centers and out of sight of anyone who is not directly involved in growing them. It's almost like we've decided that the best farm is the farm we can't see.

I'VE BEEN INTERESTED IN questions about food and health for many years. My last book, *ContamiNation*, examined similar questions about the toxic chemicals found in everyday consumer products. Big-box stores are full of things—mattresses, air fresheners, paints, cosmetics—made from some 80,000 different petrochemicals, and of these, the Environmental Protection Agency (EPA) has a full set of toxicity information for just 7 percent. Despite frightening spikes in everything from cancer rates to autism, endocrine problems and neurological disorders, 99 percent of these chemicals have never been tested for their effects on human health. In researching that book, I was shocked at the misinformation—if not the total lack of information—about the products we use in our everyday lives.

Likewise, given the amount of confusion surrounding our food system, I set out in search of facts. GMOs, and the chemicals used to grow them, have become so ubiquitous, so stitched into the fabric of our daily lives, that they are essentially invisible. To me, this invisibility is itself a problem: How can something as intimate as the food we eat be so utterly misunderstood?

My journey to find out took me from farms in New York and Maryland and Pennsylvania to plant laboratories in Delaware and Missouri and Kansas to the "ground zero" of the global GMO debate on three islands in Hawaii. During the course of my research,

I interviewed some of the world's great agricultural visionaries, some of whom take radically different approaches to the question of GMOs. One scientist, whose engineered papaya plants saved an entire industry from collapse, considers GMOs to be above reproach. Another, who is trying to invent a plant that would replace—*replace!*—millions of acres of industrial crops across the farm belt considers GMOs to be a tool the food industry has used to push the American landscape to the brink of ruin.

I spoke with brilliant farmers who think GMOs will help move the world closer to sustainability, others who think GMOs will accelerate our ecological demise. I spoke with geneticists who are developing plants that could save millions of people from starvation, or from going blind, and others who think such plants represent a Trojan horse that will do more to spread the influence of American companies than actually help the poor.

It can be hard to hold these competing stories in your mind at the same time. Clearly, genetic engineering has the potential to help solve some of the world's pressing food and nutrition problems. The problem is that this technology is mostly being used not to help small farmers or improve nutrition in the developing world but to create profits for companies selling poor-quality food in the United States. It's not *GMOs* that are a problem, in other words; it's the industrial *food system* that is the problem. That system is designed by and for the agrochemical industry to sell two enormously profitable products: chemicals, and the seeds that can withstand those chemicals. This system has been built so thoroughly around us that we don't even see it.

This book offers a look at something that is both very complex and very fundamental. Understanding what we eat, and how we have come to eat this way, requires thinking not just about food but also about history, and science, and politics, and ethics. Beneath these issues are fundamental questions of culture. How do we want to eat? How do we view the land we live on, and the plants and animals with whom we share that land? Do we trust the industries that are feeding us, or the government that is supposed to be protecting us? Do we trust that science can remain independent of corporate money and corporate power, and provide clear, independent answers to questions that directly affect our lives?

To help answer these questions, I have organized this book into three parts. Part One examines the central questions most people want to know about GMOs. Are they safe? How are they made? Are they well tested, and are the tests trustworthy? How much control does the food industry exert over government regulators? How much control does this industry have over what we are allowed to know about what we eat? More broadly, how do GMOs fit into the evolution of American culture itself, from the very small (like the birth and growth of advanced genetic science) to the very large (like the postwar development of our highways and suburbs)?

Part Two takes us to the front lines of the GMO debate to see how this system plays out—for better and worse—in real communities. On three islands in Hawaii, the battle over GMOs has been exceptionally heated, and for very different reasons. On the Big Island, a world-renowned professor created—without any help from industry—a GMO fruit that helped save the economy of his beloved homeland. On Kauai, the story is utterly different: there, a group of activists, worried about vast and secret chemical spraying used on experimental GMO farms, are fighting tooth and nail against some of the largest chemical companies in the world. And on Maui, a tiny island that nonetheless serves as the very birthplace for much of the world's GM corn, indigenous Hawaiians and local organic farmers are trying to kick the GMO industry off their island completely. For them, GMOs are not just about food, they are about the misuse of sacred land and the oppression of local people.

Part Three offers a look at alternatives to an industrial farming system that has been so destructive—and that has tarnished the reputation of GMO technology itself. I visit scientists developing GM crops they hope will prevent mass starvation in the developing world, especially as climate change threatens to undermine traditional farming practices in Africa and Asia. I spend time with farmers who use GMOs as part of a larger effort to make American agriculture more sustainable. I interview researchers who say nibbling around the edges of industrial farming isn't enough—they want to develop crops that will overthrow the entire system itself. And I speak to organic farmers—in the country, in the suburbs, and in the city—who say that no technology, no matter how exquisitely designed, will ever take the place of local people growing food for their own neighbors. Their model, they say, is the way farming was done for 10,000 years, and that GMOs, while perhaps helpful, will be useful only if they augment traditional farming practices that take seriously the health of people as well as the health of our planet.

Throughout the process of writing this book, I also tried an experiment of my own. I required my college students to wrestle with the GMO debate, and—at the same time—to work on a very small organic farm. Every week, my students explored the complexities of the American food system, and they tossed hay, fed sheep, and harvested tomatoes. They argued about the best way to feed the world, and the best way to feed themselves. Some left the conversation convinced that GMOs should have a firm place in the future of food. Scientists who can figure out a way to make drought-resistant crops that will support billions of people in a warming world deserve nothing less than a Nobel Prize.

Others were more cynical. Companies touting the benefits of GMOs are engaged in a global sleight of hand: they claim they want to feed the world, then turn around and sell us all chicken nuggets, cheeseburgers, and sixty-four-ounce sodas.

Other students were more philosophical. They left convinced that the primary problem in the American diet is not nutrition or any particular technology, but ignorance. If one way to improve the way we eat is through fancy new technology, maybe another way is to get more people—including English majors—to spend time working on local, small-scale, organic farms. Getting "more eyes on the acre," they said, may be the only way to close the enormous gap that has opened up between most American people and the food they consume every day.

To my mind, they were all right. I hope the following chapters will explain why.

Part One

ROOTS

Are GMOs Safe?
Is That the Right Question?

The first thing, and sometimes the only thing, that people want to know about GMOs is simple: Are they safe to eat? It's an obvious question, since we're all consuming them at almost every meal, and a legitimate one, since it's not always clear what GMOs are, how they are made, or where they appear in our diet. In the decades since the creation of the Flavr Savr tomato, we are all eating genetically modified food, whether we know it or not, and whether we like it or not.

Add to this the fact that basic information—even in the form of simple labels on food—is very hard to come by. Although you most likely eat GMOs for breakfast, lunch, and dinner, there is simply no way to know it.

Most people, sitting down for a meal, would rather not wrestle with the way small RNAs affect the chromosomes of the corn that went into the cow that went into the burger they are eating. They would certainly rather not contemplate whether that same corn had something to do with climate change, or the obesity epidemic, or the decline of bee populations, or whether they contribute to water pollution, the pesticide contamination of our bodies, or the destruction of small-town America.

In a way, asking whether GMOs are "safe" is like asking whether Froot Loops are safe, or cheeseburgers, or nail polish: if you narrow the question down enough, the answer is almost certainly, sure, GMOs are "safe," but "safe" may not be the same thing as "good for you." Not many people get sick from eating a single bowl of Froot Loops, and not many people get sick from painting their nails once or twice. But how many bowls, or manicures, would it take to make a product "unsafe"? A great many molecular biologists argue that altering a plant's genetic structure simply mimics natural evolutionary processes and that GM foods are more fully studied—and at least as safe to eat—as anything has ever been. Many hundreds of studies have supported this: food made from plants that have been genetically engineered do not appear to be any more harmful than food grown traditionally.

Those who create, control, and profit from GMOs—the scientists who develop and use the technology, along with the biotech and food companies that make up our industrial food system—consider the debate over genetic engineering to be fully settled. So do highly reputable scientific organizations. The science of GMOs is clear, they say: the technology has been around for decades and has developed into a highly precise method of producing enough food to feed the earth's 7 billion people.

Genetic engineering is simply an incremental step—a new tool, geneticists and molecular biologists say—in the long progression of agricultural science. Since the dawn of the agricultural era 10,000 years ago, farmers have selected seeds from the season's most successful crops and discarded the seeds from the least. This "human selection" is merely a manipulated version of the "natural selection" that forms the bedrock of evolution. Tinkering with a plant's genome is no different from evolutionary processes that have gone on since time began. Faster, perhaps, but no different.

These techniques are no more risky than induced mutation, or "mutagenesis," the long-standing practice of exposing seeds to chemicals or radiation to induce random mutations. Mutations happen all the time in nature, and some produce plants with favorable traits like drought tolerance, or higher yields, or better taste. In the last century, more than 3,200 mutagenic plants—from pears to peanuts, from barley to grapefruit—have been released on the market. These crops are not GMOs, and they are considered so benign they are even allowed on organic farms.

But the minute you open the aperture a bit, the question of "safety" becomes considerably more complicated. While the *process* of engineering plants may be considered "safe," the *consequences* that ripple out from it are considerably more troubling. The molecular structure of a single GM plant may not be a cause for alarm, but what if almost all GM crops are grown to produce things like cheeseburgers and salty snacks and soft drinks, which have ramped up the country's obesity epidemic? Is that a GMO problem, or not?

What about the chemical pesticides and herbicides—many of them known to cause both health and environmental problems— that are sprayed on hundreds of millions of acres of GM crops? These chemicals existed long before GMOs, of course; indeed, they were developed decades ago by the same companies (Monsanto, DuPont, Dow, Syngenta) that are now the world's leading sellers of GM seeds. Critics often say that GMOs are less necessary for making food than they are a powerful vehicle for selling pesticides; once a company has sold farmers on the idea of GM seeds, they are far more likely to buy chemical sprays that go along with them. If they were using the company's chemicals already, why not also buy seeds that are resistant?

So are pesticides a "GMO problem," or are GMOs just exacerbating the problem of industrial farming itself?

More broadly, what happens when entire global industries—and entire swaths of North America—are constructed to keep cheeseburgers and snacks and soft drinks (and thus the GMOs that make them) flowing into our bellies? What if these industries become so enormously profitable, heavily marketed, and politically powerful that the foods they produce began to seem "conventional" (or stranger still, "traditional")? If problems—even deep problems—began to crop up, would we even be able to see them?

In other words, most people involved in the GMO debate—no matter what side they are on or how passionately they argue their position—consider the narrow question of safety to be the wrong question.

"I've been a lawyer for over thirty years, and this is by far the most polarized issue I've ever dealt with," Paul Achitoff told me. Achitoff is an environmental attorney for EarthJustice, which is handling a series of major GMO lawsuits in Hawaii. Achitoff has been in the GMO trenches since the beginning.

"Inevitably, no matter what the subject matter—pesticides, labeling—people always spend their time talking about how dangerous GMOs are to eat. All people want to know is, 'Is it healthy, is there proof?' People in favor of GMOs say they are safe as mother's milk. Others say they are dangerous. I don't even bring that subject up in court. To me, it's not even relevant. It's not even reasonably disputed that there are environmental and socioeconomic consequences here."

Indeed, a great many organic farmers, a wide swath of health, consumer, and environmental organizations, and First Amendment "right to know" advocates say the GMO debate is about a lot more than molecular science. GMOs, in this view, are the very symbol of all that is wrong with the American food system. Whether or not the technology involved in genetic engineering is "safe" (and not all opponents are willing to concede this point), the crops—along with the pesticides and herbicides used to grow them—represent a profound insult to public health and ecological balance.

"Nature's been around a long time, so to think we can dance in there and take a gene off the shelf and get a product that your body will accept is really arrogant," Gerry Herbert, an organic farmer and anti-GMO activist in Hawaii told me. "It's like throwing a wrench in a moving engine. You're going to have a problem. We don't even know what's in the soil, and yet we're killing it because we can get a quick profit from it. Do we want corporations to control our food? Their whole mandate is to maximize profit. That's why they're there. Are they worried about your nutrition? Not a bit. They will do everything necessary to rearrange genes to maximize their profits."

Regardless of its effect on a plant's molecular structure (or that plant's impact on our bodies, or the ecosystem of which the plant is a part), GMO technology is mostly used to turbocharge the engines of an unsustainable farming system that is dousing our land and water with chemicals, wearing out our soil, making us fat, and lining the pockets of companies that already hold far too much economic and political power.

There is truth on both sides of this debate. There are also half-truths and naked cynicism. There are scientific studies that say GM foods are entirely safe to eat, and others that say they aren't. Earth Open Source, an organization run by the molecular biologist John Fagan, recently published a book called *GMO Myths and Truths* with more than 300 pages of studies arguing that GMOs are unhealthy for our bodies, our environment, and our political and economic systems. "GMO Answers," a website overseen by the biotech industry, is larded with studies heralding the benefits (and safety) of GMOs, as well as essays designed to make you feel better about your own doubts ("Skeptical About GMOs? We Understand.").

Which side are we to believe? Consumers can be forgiven for feeling that questions about safety ought to be simple: Is GM food safe to eat, or not? The trouble is, there are complexities at both the micro and macro levels that make such questions of "safety" a lot more complicated than they might first appear. The few journalists who have tried to navigate this jungle have found themselves with few reliable guideposts.

"The quest for greater certainty on genetic engineering leaves you chasing shadows," noted Nathanael Johnson in the magazine *Grist*. "When you're dealing with gaps in knowledge, rather than hard data, it's hard to tell what's an outlandish hypothetical and what's the legitimate danger. Anything, of course, is possible, but we shouldn't be paralyzed by unknown risks, or we'll end up huddled in our basements wearing tinfoil hats."

So let's take a closer look at this.

FARMERS HAVE SPENT countless generations crossbreeding, or hybridizing, closely related plants to create desirable traits in their offspring, like bigger fruit, higher yields, and better taste. Do this over and over for many generations and you end up with the apples and lettuces and carrots we recognize today.

Genetic engineering, in this line of thinking, is nothing more than human selection, sped up. GM plants are of two varieties. They are either "cisgenic," which means they are created by taking a gene from a wild apple tree, for example, and stitching it into the genome of a domesticated apple, to prevent the fruit from scabbing. Or they are "transgenic," meaning they are created by taking a gene from one kind of organism (a bacterium, for example) and inserting it into the genome of another kind of organism (a corn plant, say) to help make the corn resistant to plant-eating insects.

There are only four kinds of genetically engineered plants currently approved for agricultural use: those (like Roundup Ready soybeans) that tolerate the herbicides farmers use to kill weeds; those (like Bt corn) that are engineered to produce their own insecticide; those (like Plenish soybeans) made with altered nutritional components, like healthier fatty acids; and those (like most papaya grown in Hawaii) that have built-in virus resistance. Many other potential applications are in various stages of development.

While nothing is absolutely certain when it comes to the interplay between food and health, it seems fair to say that one claim made by industry and its scientific allies is correct: Every day, hundreds of millions of people, in twenty-eight countries, eat food made from (or eat animals fed from) GM plants. Many scientists are willing to leave it at that. After billions of meals served with GM ingredients, "no adverse health effects attributed to genetic engineering have been documented in the human population," the National Research Council and Institute of Medicine say. The American Academy for the Advancement of Science agrees: "Contrary to popular misconceptions, GM crops are the most extensively tested crops ever added to our food supply." The World Health Organization considers GMOs to "have passed risk assessments in several countries and are not likely, nor have been shown, to present risks for human health." The scientific adviser to the European Commission has said, "There is no more risk in eating GMO food than eating conventionally farmed food."

Most GMO studies have been done on animals, which makes sense, since food-producing animals consume as much as 90 percent of the GM crops grown worldwide. In the United States,

95 percent of the 9 billion cows, hogs, chickens, and turkeys raised for food eat GM grains. A recent meta-analysis of studies looking at some 100 billion livestock animals raised between 1983 (before the introduction of GMOs) and 2011 (long afterwards) found no "unfavorable or perturbed trends" in animal health or productivity. "No study has revealed any differences in the nutritional profile of animal products derived from [GMO]-fed animals," reported researchers at UC-Davis.

According to Blake Meyers, a plant geneticist at the University of Delaware, genetically altered plants have been so thoroughly studied that the question of whether or not they are safe to eat is no longer even an interesting scientific question. "We can say that these products and genes are as safe as we can know, and thus far, the track record of GM products has shown that they are safe," Meyers said. "The study of GM food products already approved for commercial use isn't a topic of interest to most plant biologists/scientists because the interesting work on them was done years ago, and they are so exhaustively studied that you'd have to work really hard to find something new."

Sure, there remain gaps in our knowledge about genetic engineering, Meyers says; the field is still only a couple of decades old, and new discoveries about how plants function are happening all the time. But in terms of safety, genetically modified products "are very well characterized, so I would say that by the time they're taken to market, they're extraordinarily well tested, and they're both predictable and reliable. GM products are also exhaustively analyzed, much more so than nontransgenic food products, so the possibility of an important gap in our knowledge about the introduced genes is typically extremely small."

Anxiety over GMO technology has more to do with the human

fear of the unknown than it does with actual risk, said Jim Carrington, a plant pathologist. Carrington's credentials are impressive: he's a member of the National Academy of Sciences and the president of the Donald Danforth Plant Science Center, one of the leading nonprofit plant research centers in the world. "Do we really have so much knowledge about small RNAs or the impact of adding a single gene or two through a GMO approach—do we know so much that we can eliminate any risk? The answer is clearly no," Carrington told me. "But no approach is risk free. We do not have the ability to be confident that we have eliminated all risks. That is the basic fact of risk assessments: you can do your best to assess impacts based on data, and you can know with a high degree of confidence that risks are relatively low and worth taking in view of the benefits.

"But anyone who says the aim should be to wait until all the data are in, that's foolish. All the data will never be in."

The question about GMOs, Carrington said, should not be "Are there risks?" but "What does science tell us about what is the reasonable likelihood of a problem coming to bear?" In the case of GMOs, "the science has been pretty clear. There are over a thousand journal articles that collectively say that the risks are exceedingly low from the standpoint of comparisons to all alternatives—conventional or organic agriculture. The risk is simply very, very low."

But as confident as Meyers and Carrington are—and they represent the majority of scientists working with GMOs—their opinions are not universal. The trouble with such proclamations, critics say, is that genes don't function as neatly (or as predictably) in the world as they do in the laboratory. Instead, they function in the enormously subtle context of other genes (within the organism itself), other organisms (in the soil and in the creatures that eat them), and other ecosystems (in the world at large). There is a randomness in genetics, an unpredictability that lies at the heart of reproduction, and it is this imprecise nature of genetics that scientific critics of GMOs frequently invoke as reason for caution.

A genome itself is a kind of microscopic ecosystem, and "we all know what can happen when you, for example, try and introduce a single species into an ecosystem," John Vandermeer, an ecologist and evolutionary biologist at the University of Michigan, has written. "What usually happens is nothing, which of course can lead to complacency. But occasionally the introduction is catastrophic."

Cane toads in Australia, Nile perch in Africa, kudzu in the American South—there are countless examples of ecological disasters caused by introduced species, Vandermeer writes. "If genomes are like ecosystems, there is nothing at all that suggests equivalent disruptions could not occur, and the few scientists who remain unaware of this complication need to refresh their graduate education with a course in complex systems."

And inside our bodies? One of the most frequently raised concerns about GM foods has to do with toxins and allergies. GMOs can introduce proteins into our diet that the human body has never encountered before, and food allergies seem to be rising everywhere. While evidence of a direct link is scarce, the long-term effects of eating clinically undetectable traces of new proteins remain a concern.

Alfredo Huerta, a plant biologist at Miami University in Ohio, pointed me to a short-term (thirty-one-day) study that showed that eating GM corn causes abnormalities in the digestive systems of pigs. A two-year study of pigs fed a mixture of GM corn found they developed severe stomach inflammation (and 25 percent heavier uteruses) than pigs fed non-GM corn. The findings were troubling for a couple of reasons. First, pigs have digestion systems similar to those in humans. Second, the pigs were sickened not by a single GM grain, but by a mixture of different GM grains. Mixed grains, the authors noted, are not tested for toxicity by regulators "anywhere in the world."

As for humans? In his biology classes, Huerta tells his students that he will give an A to anyone who can show him a long-term clinical trial in humans showing that GMOs are safe.

No one has ever found one.

When industries say that GMOs are safe because billions of people have eaten them and no one has dropped dead, they're being anecdotal, not scientific, Huerta told me. How would we even know if large-scale physical symptoms are caused by GMOs if we don't even know we're eating GMOs? Even leaving aside major issues like cancer or endocrine problems, how many other symptoms—headaches, stomachaches, allergic reactions, changes in the way our immune system functions, microscopic changes in the structure and function of our cells—may be caused by GMOs if we don't know where these ingredients enter our diet, and if we don't conduct proper human clinical trials?

"We tend to blow off the reason for a migraine, the ill feeling that we had, on something that we will never be able to identify," Huerta said. "How do we know if any of those hidden symptoms are due to having consumed a GMO (such as GM sweet corn, which is designed to be eaten fresh, right off the cob, and full of Bt toxin)? Remember that physical ailments due to smoking usually appear after many years. Things like emphysema, asthma, loss of lung function, secondary metabolic effects, etc. tend to show up after many years of smoking. Do we know if anything like that will happen with GMOs? The answer is no. We don't know the answer to that question."

Huerta's skepticism is well founded. Although it is virtually impossible to lay a single illness, let alone an epidemic, at the feet of a single product, that doesn't mean these GMOs are *not* causing problems. It may just mean that we haven't made the connection yet. These foods are a new thing on the evolutionary scene, and we are eating them in unimaginably vast quantities. While it is true that most scientific research done to date has found little reason to worry, there are other truths (as we will see) that ought to give us pause.

"The fact is, it is virtually impossible to even conceive of a testing procedure to assess the health effects of genetically engineered foods when introduced to the food chain," said Dr. Richard Lacey, a member of the British Royal College of Pathologists. "The only way to base the claims about the safety of genetically engineered food in science is to establish each one to be safe through standard scientific procedures, not through assumptions that reflect more wishful thinking than hard fact."

Is It the GMOs, or the Chemicals We Spray on GMOs?

One health concern about which there is considerably less doubt is that GMOs, from the very outset, have been developed alongside synthetic pesticides and herbicides. The companies that sell the most GM seeds—Monsanto, DuPont, Dow, Syngenta—all started out as chemical companies, and their move into the seed business, whatever else it has done, has vastly expanded their capacity to sell chemical sprays.

Even the most benign of these chemicals are known to cause health and environmental problems, and they are used in enormous quantities. In the United States over the last forty years, the use of glyphosate (sold by Monsanto as Roundup, a product that makes the company $5 billion a year) has grown by a factor of 250, from less than half a million to 113 million kilograms a year. It is so common in England that residues of the compound routinely show up in British bread. A study by David Mortensen, a plant ecologist at Pennsylvania State University, predicts that total herbicide use in the United States will double again before 2025 as a direct result of GM crop use.

Glyphosate has been approved by the EPA and regulatory agencies all over the world, and has earned the lasting loyalty of countless farmers who use it to clear fields of weeds. Scores of studies have shown no link to cancer; a recent report by the German Federal Institute for Risk Assessment found that glyphosate is not carcinogenic or toxic to fertility in lab animals.

But this opinion is far from unanimous. The International Agency for Research on Cancer—the cancer research arm of the World Health Organization—recently declared that glyphosate and 2,4-D (another common herbicide) should be classified as, respectively, "probable" and "possible" human carcinogens. France, the Netherlands, and Sweden have all recently come out against relicensing glyphosate for use in the European Union.

Within American regulatory agencies, scientists have long been troubled by the influence industry holds over government regulators. There is a well-documented pipeline leading from industry employees to EPA staff, and industry lobbyists have been very effective at limiting federal funding for chemical regulation. Until the summer of 2016, the federal Toxic Substances Control Act, the government's primary tool to regulate chemicals, had not been updated in forty years. In the early 1970s, there were a dozen EPA laboratories dedicated to testing farm chemicals. In 2004, thanks to decades of industry-pressured "deregulation," there were two.

Chemical companies routinely hire former senior government officials to help them design corporate strategies and to persuade their former colleagues in government to be lenient in their scrutiny of data. And they are adept at getting their own people into positions of power in government. This was most obvious during the Reagan and George H. W. Bush years, when "regulatory relief" led to a dramatic dismantling of the EPA—and such breaches of the public trust by former industry insiders that several were forced to resign for ethics violations and one even went to prison.

President George H. W. Bush appointed Clarence Thomas, a former lawyer for Monsanto, to the Supreme Court; Thomas later wrote the majority opinion in a landmark case granting companies the right to patent GMO seeds. In the 1990s, President Clinton got so cozy with Monsanto's CEO Robert Shapiro that he swooned over the company in his 1997 State of the Union address and named Shapiro to the president's Advisory Committee for Trade Policy and Negotiations. There, Shapiro worked closely with Mickey Kantor, Clinton's trade representative before becoming a Monsanto board member himself. In 1998, Clinton personally awarded the National Medal of Technology and Innovation to the Monsanto team that invented Roundup Ready soybeans.

During the Obama administration, Michael Taylor, a former Monsanto vice president, was given a senior position in charge of food safety at the FDA. Islam Siddiqui, a Monsanto lobbyist, was named the U.S. Agricultural Trade Representative, put in charge of promoting American farm products overseas. As Obama's U.S. Solicitor General, Elena Kagan wrote a brief requesting the Supreme Court lift a ruling forbidding the planting of Monsanto's genetically engineered Roundup Ready alfalfa. Kagan now sits alongside Clarence Thomas on the Supreme Court.

"From the 1940s to the dawn of the twenty-first century, it has seemed as if government has been working for industry rather than overseeing it," E. G. Vallianatos, a twenty-five-year veteran of the EPA's Office of Pesticide Programs, has written. "Most government and academic scientists working on agricultural practices and pest control have obdurately ignored research into nature's intricate and subtle workings. Instead, they have smoothed the way for the poisonous (and hugely profitable) concoctions of the chemical industry, and they are now doing the same for the rapidly growing field of genetic crop engineering."

There will be more on this later in the book, but suffice it to say that the debate over the safety of farm chemicals, like the debate over the safety of GMOs themselves, remains fractious and tangled up as much in money and politics as in concerns for human health.

Food companies like to say that GMOs have reduced the total load of chemicals sprayed on crops, and in one way this is true. Between 1996 and 2011—the first sixteen years of broad GMO planting—the use of the insect-resistant Bt crops (plants inserted with genes from a naturally occurring bacteria found in the soil) reduced the use of insecticides by 123 million pounds. But during those same years, the use of weed killers like glyphosate and atrazine rose by 527 million pounds.

The net result? An increase of 7 percent, 404 million pounds. Part of this, at least, is the result of a chemical feedback loop: the more farmers use sprays, the more weeds evolve resistance to sprays, which means farmers need to use more, and stronger, chemicals. The magnitude of the increase in herbicide use on GM crops has "dwarfed" the reduction in insecticides used on Bt crops, the agricultural economist Charles Benbrook reports, "and will continue to do so for the foreseeable future."

No matter how you slice it, that's a lot of synthetic chemicals going onto (and into) our food. David Pimentel, a Cornell University scientist who has been studying American agriculture for fifty years, has estimated that pesticides cause some 300,000 poisonings a year in the United States; worldwide, the number is more than 26 million, 3 million of whom required hospitalization. Every year, pesticides kill 220,000 people worldwide and cause chronic illness— everything from respiratory problems in farmworkers to cancer and hormone problems in consumers—in another 750,000.

"The majority of food purchased in supermarkets have detectable levels of pesticide residue," Pimentel writes. In 1982, 80 percent of the milk supply on the Hawaiian island of Oahu had to be destroyed because it had been contaminated with the insecticide heptachlor. But at least heptachlor was on the regulatory radar: of the six hundred pesticides now in use, federal regulators search for the residues of only about forty.

And those numbers tabulate just the risks for humans. Pimentel has also found that agrochemicals kill some 70 million birds every year in the United States alone. A quarter-million domestic animals are also poisoned every year by pesticides; farmers lose some $30 million a year to animal illness and death caused by pesticide poisonings—an estimate considered low because it includes only numbers reported by veterinarians. "When a farm animal poisoning occurs and little can be done for the animal, the farmer seldom calls a veterinarian but, rather, either waits for the animal to recover or destroys it," Pimentel writes.

It is true that pesticides and herbicides are not GMOs, and it is also true that farmers sprayed all kinds of chemicals on their crops long before the development of GMOs. Consider wheat, which is not (currently) genetically engineered. Wheat is often sprayed with glyphosate as a desiccant immediately before it is harvested, in order to force the plants to rapidly release their seeds. This puts a concentrated chemical on the plant right before it's processed into food.

Some scientists wonder whether the rash of gluten intolerance currently afflicting the nation is actually Roundup intolerance. Glyphosate may be "the most important causal factor" in celiac disease, one study recently fou d; another found that glyphosate exposure can cause severe dep! ic of the utrient manganese, a deficiency of which is associa ev..ything from anxiety to autism. "The monitoring of glyphosate levels in food and in human urine and blood has been inadequate," the study's authors reported. "The common practice of desiccation and/or ripening with glyphosate right before the harvest ensures that glyphosate residues are present in our food supply." It is also plausible that "the recent sharp increase of kidney failure in agricultural workers is tied to glyphosate exposure."

This, then, is not a question of "the safety of GMOs"; it is a question of "the safety of what we spray on our food," a whole lot of which *is* GM. It's obviously impossible to pin a nation's health woes on a single chemical compound, especially when only a tiny fraction of the country's 80,000 synthetic compounds have ever been formally tested for their health consequences. However, few chemicals have been spread as far and wide as glyphosate in the last twenty years, and glyphosate's ability to disrupt the body's detoxification pathways has been shown to intensify the effect of other toxic chemicals.

There is no question that the explosion in the use of chemicals like glyphosate has tracked right alongside the explosion in the use of GMOs. It has also corresponded with two other trends: a "huge increase" in the incidence and prevalence of chronic diseases, and a "marked decrease" in life expectancy in the United States, write the authors of a study published in *The Journal of Organ Systems*. Diabetes, obesity, cardiovascular disease, neurological diseases—all have jumped dramatically, to the point that one-quarter of Americans now suffer from multiple chronic diseases. These numbers run parallel to "an exponential increase in the amount of glyphosate applied to food crops and in the percentage of GE food crops planted." The annual cost of treating these illnesses, according to the Centers for Disease Control and Prevention (CDC), is north of $750 billion per year. While direct, causal links have not been firmly established, the correlation—especially given the scale of our exposure to pesticides, GMOs, and the processed foods both help create—surely merits more attention than it has thus far received.

Beyond worries about human health, there is the question of how much longer the pesticides currently associated with GMOs will remain viable. Farmers have sprayed so much glyphosate on their GM crops that weeds—the very things they use glyphosate to control—are evolving resistance to the spray. In 2004, a common weed called amaranth was found to have developed resistance to glyphosate in a single county in Georgia; by 2011, it had spread to seventy-six. "It got to the point where some farmers were losing half their cotton fields to the weed," a Georgia farming consultant reported.

Glyphosate-resistant weeds have now been found in eighteen countries, with significant impacts in Brazil, Australia, Argentina, and Paraguay. In the United States, they have emerged on 100 million acres in thirty-six states, meaning farmers must now return to harsher chemicals (like atrazine, a known carcinogen) or to recently approved "stacked" herbicides that combine glyphosate with 2,4-D, a component of Agent Orange, the carcinogenic defoliant used during the Vietnam War. Dow AgroSciences, which uses 2,4-D in an herbicide called Enlist Duo, says there are more than 1,500 products with 2,4-D as an active ingredient; over the next few years, the EPA predicts the use of 2,4-D will increase sevenfold.

The EPA's recent decision to approve "stacked" herbicides was deeply flawed, according to Philip Landrigan, a renowned pediatrician and public health scientist, and Charles Benbrook, an agricultural economist at Washington State University. The decision was based on studies done thirty years ago, which were not done by independent scientists but by the herbicide manufacturers themselves and were never published. The EPA did not take into account what scientists now know of the dangers such chemicals pose—even at very low doses—to the human endocrine system, especially in children. And they failed to consider the chemicals' impact on the environment, especially its effect on pollinators like the monarch butterfly, whose population is down more than 90 percent.

By pushing chemicals like glyphosate so hard, and for so long, chemical and seed companies "have sown the seeds of their own destruction," the University of Michigan's John Vandermeer told me. "We now have twenty-five weeds that are Roundup resistant, so now they're developing 2,4-D–resistant crops. Roundup and 2,4-D are not good things to have around in such huge quantities. Roundup is toxic to amphibians—it's actually toxic to almost everything that people have studied.

"My worry is that spreading Roundup all over the place has not been a good idea, and now we're about to start spreading 2,4-D around the world. It's not a good idea for the environment, and it's a potential danger for human health. Both chemicals are certainly suspected carcinogens, and Roundup is an endocrine disrupter. These are problems that were well known before there were GMOs. I don't care what technique you use to create Roundup Ready crops. I will always have an objection to the chemicals they encourage. If they had created Roundup Ready crops the old-fashioned way, I still wouldn't like them because of the Roundup."

It would seem that with so much riding on this question of safety—with so much food, so much health, and so much money riding on a clear answer—the federal government would make answering it a priority. The trouble with federal oversight of widely used chemicals like glyphosate is that the agencies responsible for keeping an eye on industry are deeply compromised by the political power of these same industries. The EPA has "gutted" both internal and external research programs responsible for safeguarding the public from industrial and agricultural chemicals, Bruce Blumberg, a professor of developmental and cell biology at the University of California, Irvine, told me. The EPA says everything they do is online, but "damned if you can find it," he said.

Relying on the seed and chemical companies to test their own products is folly, Blumberg said. Especially for something as ubiquitous as Roundup, large, long-term, and multigenerational studies ought to be carried out by a federal agency like the National Toxicology Program.

"This kind of work is the province of government, but they have totally shirked their responsibility," Blumberg said. "We just cannot trust people with financial interest in product sales to do safety tests on these products. Companies will never show all the data unless it is in their interest. Look at history. Look at the tobacco industry. Look at General Motors and the ignition switch debacle at Takata and their exploding air bags. What does history tell us? Nothing good."

The Information Squeeze

It is the absence—or, if you like, the impossibility—of an absolute proof of safety that has led more than sixty countries all over the world to require foods containing GMOs to be labeled. With certainty so hard to come by, these countries (notably not including the United States) have decided that consumers at least deserve enough information to decide what they want to eat.

Europeans have bitterly opposed GMOs since the beginning. Their objections cropped up right around the time people in England learned that cows were being fed the brains of other cows. Mad cow disease, which had nothing to do with GMOs, nonetheless made people skittish over both the excesses of industrial agriculture and the paucity of government regulation.

Theatrical demonstrations popped up all over Europe and quickly focused on agricultural technologies of all kinds. Protesters dumped GM soybeans at the doorstep of the British prime minister. Food activists pressured supermarkets to pull GMOs off their shelves. Prince Charles said GM foods took mankind into "realms that belong to God."

In 1996, the German division of Unilever canceled an order for 650,000 metric tons of soybeans unless they could be guaranteed not to contain GM beans. Four years later, the EU required that food with more than 1 percent GM ingredients carry a label. Such was the European resistance to GMOs that hardly any foods ever actually ended up with a label, because hardly any GM foods were actually available for sale. Around this time, a food analyst for Deutsche Bank in New York declared that "GMOs are dead."

More recently, nineteen members of the European Union requested that they be able to "opt out" of an agreement that allows the planting of GM corn.

In the United States, poll after poll indicates that a majority of people are confused and frightened by engineered food, and that they share a deep mistrust of the large agribusinesses that make them. They worry about the evolution of superbugs and superweeds, and about the growing dangers of pesticides and synthetic fertilizers needed to keep industrial farms afloat. They worry about the creation of dangerous food allergies, like a GM soybean made with genes from a Brazil nut that became such a suspicious cause of allergies that it never made it to market.

"The GMO issue is something that continues to be brought up in an unprompted way in our interviews with consumers," said Laurie Demeritt, CEO of the food research firm Hartman Group. "And when we look at things like fat, sodium, and sugar, GMO is showing the strongest growth rate in terms of characteristics that consumers are trying to avoid. . . . Consumers have a vision in their minds of people in lab coats taking syringes and injecting things into a product, a vision of food made in a lab—and that's even worse in their minds than food coming off a factory line."

In a 2013 *New York Times* poll, three-quarters of Americans surveyed expressed concern about GMOs in their food, with most worried about health risks. More than 90 percent of Americans want GMOs labeled, as they have been required to be in countries such as India, China, Australia, and Brazil.

In 2011, Gary Hirshberg, chairman and cofounder of Stonyfield, the organic yogurt company, partnered with Just Label It, a national coalition of nearly 450 organizations, to petition the FDA to make GM food labels mandatory. More than a million people have now signed up. In 2014, Vermont became the first state to require labels on foods made with GMOs (though critics complain that the state left a sizable loophole by exempting meat and dairy products, much of which comes from animals fed GM grain).

Companies have responded aggressively to these moves. They have spent tens of millions of dollars in the United States alone trying to limit the information they must provide about the GM ingredients in their food, or the pesticides they use. They fight citizen groups at the ballot box and pour rivers of money into the pockets of politicians who support them. They place industry insiders at the very top of the federal agencies charged with regulating their own industry. They invest millions of dollars in university laboratories, then urge the scientists they support—who the companies know "have a big white hat in this debate"—to explain the benefits of their products in the press and before Congress.

When California activists decided to float a petition for food labeling in 2012, they gathered more than a million signatures (and $9 million) in support of Proposition 37. The move was derailed by a massive counter-campaign (and $46 million) from Monsanto, DuPont, Pepsi, and Kraft Foods. In the end, the labeling measure failed 51 percent to 49 percent.

The story repeated itself in Oregon and Washington state: small-scale activists in favor of labeling followed by multimillion-dollar campaigns financed by the food and agricultural industries. "Monsanto was writing million-dollar checks at a shot," recalled Trudy Bialic, the public-affairs director of a Seattle-based natural-foods co-op chain, who helped draft the initiative. The Grocery Manufacturers Association, the lobby for makers of processed food, donated $11 million. "Boom, boom, boom, millions overnight," she said. "It was death by a thousand cuts."

If some of this sounds familiar, it should. Companies pushing the "safety" of GMOs are following a playbook written by Big Tobacco and Big Oil, which spent decades claiming that science (about cancer, or about climate change) was bunk. Yet now, when consumers demand to know more about GMOs—what they are,

how they are made, what their health and environmental consequences might be—industry claims to have science "on its side."
Consumers should trust these companies to do the right thing, because the science on GMOs is "clear." According to a recent survey
by the Pew Research Center and the American Association for the
Advancement of Science, the gap between what scientists and the
public believe about GMOs is now wider than on any other issue.
Almost 90 percent of scientists believe eating GMOs is safe. Among
the public, that number is 37 percent.

The food industry has always been heavy-handed in its battles
over what information the public should be allowed to know. Twenty-
five years ago, many of these same companies fought bitterly to prevent the legally strict "organic" label from being applied to foods
grown without synthetic chemicals. They had reason to be concerned: since the introduction of organic standards, the organic
food industry has been growing at 20 percent a year, which has both
cut into traditional profit centers and opened the door to a whole
new array of growers, preparers, and sellers of food.

But the GMO labeling debate has a different feel. Requiring
a "Contains GMOs" label on foods would function as a kind of
"anti-organic" label, implying (given the public's anxiety over the
issue) that the food was somehow unsafe to eat. To big food companies and farmers who use GMOs, requiring a GMO label would do
little more than give the organic food industry another big bite out
of the American food budget.

As consumer anxiety over GMOs has grown, so have the marketing opportunities for food companies that do not use GMOs.
Some 80 percent of consumers say they would pay more for foods
carrying a "No GMO" label, even though they don't necessarily
trust food labels (or even fully understand GMOs). Whole Foods

has pledged that by 2018 it will replace some foods containing genetically modified ingredients and require labels on others. Signs in Trader Joe's proclaim: "No GMOs Sold Here." Sales of products claiming they contain "no GMOs" exceeded $10 billion last year and grew at a faster rate than sales of gluten-free items, according to a recent Nielsen study.

"There's no doubt that the industry is fighting a rear-guard action on this and trying to put it to rest," said Carl Jorgensen, director of global consumer strategy for wellness at Daymon Worldwide, a consumer research and consulting firm. "But there's an aura of inevitability about it now."

Ten billion dollars for non-GMO foods is a lot, but it's still a vanishing fraction of the $620 billion Americans spent in grocery stores in 2013. But if a traditional grocery chain like Kroger or Safeway were to begin labeling its private-label products, "that would be a game changer," Jorgensen said. Unlike food manufacturers, grocery stores interact directly with consumers, Jorgenson noted; they can see which foods fly off the shelves and which foods remain.

But this is tricky magic: if companies start boasting that some of their products (like Cheerios) do not contain GMOs, how will consumers react to their other products (like Lucky Charms and even Honey Nut Cheerios)—sitting right there on the same shelf—that do?

In the absence of broad labeling laws—there are currently eighty-four bills on GMO labeling in thirty states—companies hoping to take advantage of GMO anxiety have found other solutions. On its website, a testing organization called the Non-GMO Project—logo: monarch butterfly—shows a photo of a little blond girl carrying a sign saying "I Am Not a Science Experiment."

"The sad truth is many of the foods that are most popular with children contain GMOs," the site reports. "Cereals, snack bars, snack boxes, cookies, processed lunch meats, and crackers all contain large amounts of high-risk food ingredients. In North America, over 80% of our food contains GMOs. If you are not buying foods that are Non-GMO Project Verified, most likely GMOs are present at breakfast, lunch, and dinner."

The Non-GMO Project, which calls itself "North America's only third-party verification for products produced according to the rigorous best practices for GMO avoidance," says it has verified more than 34,000 products. The nonprofit group tests ingredients, and anything passing the European standard of less than 0.9 percent GMO is eligible for the "Non-GMO Project Verified" seal of approval. Given the almost unavoidable reality of seed and crop cross-contamination, getting to zero—getting to actually "GMO free" is (so far) impossible.

Rather than mandatory labels on products, the food industry has long pushed the use of voluntary QR barcodes on products, which (they say) consumers could simply scan with their cell phones. The codes would direct you to the company's website, which would reveal further information about the product. The U.S. Secretary of Agriculture has said the QR codes would solve the label debate "in a heartbeat."

Pro-labeling groups consider this move a joke. Bar codes directing you to the Internet make abstract what ought to be utterly present and clear: Does the package in your hand contain GMO ingredients, or not? If you actually take the time to navigate to a company's website, you might (perhaps) find somewhere (in small print) that yes, Coca-Cola uses GM corn to make its high fructose corn syrup; or yes, children's breakfast cereals are sweetened with crystals made from GM sugar beets; or yes, Crisco oil uses GM soybeans. But who's actually going to go to all that trouble? Add to this the fact that 50 percent of the country's poor and 65 percent of the elderly do not even own smartphones, and you have to wonder: Is the goal of this move broad public awareness of what goes into food, or another way for companies to obscure what they are feeding us?

As with the regulation of toxic chemicals in products like cosmetics or baby bottles, companies have also worked hard to limit the size of their battlefield: a single piece of legislation in Congress is a lot easier to manipulate than bills passing through dozens of state legislatures. In 2014, in the midst of major GMO labeling battles in places like California and Vermont, Rep. Mike Pompeo (R-Kansas) introduced a federal bill seeking to prohibit states from requiring GMO labels on food. Opponents of the measure dubbed it the "DARK" Act, for "Deny Americans the Right to Know," and hundreds of thousands of people signed petitions opposing the bill. "If the DARK Act becomes law, a veil of secrecy will cloak ingredients, leaving consumers with no way to know what's in their food," said Scott Faber, senior vice-president of government affairs for the Environmental Working Group. "Consumers in sixty-four countries, including Saudi Arabia and China, have the right to know if their food contains GMOs. Why shouldn't Americans have the same right?"

Opponents also considered Pompeo's bill a gift to Big Food, and indeed, the Pompeo campaign's top individual contributor has been Koch Industries Inc., the energy, agricultural chemical and fertilizer conglomerate run by billionaire brothers Charles and David H. Koch, who are known for their extensive support of conservative political causes.

In the end, Big Food won. In the summer of 2016, President

Obama signed the Senate version of Pompeo's bill into law. Although the administration pitched the move as a step forward in the march toward consumer information, the law accomplished most of what Big Food desired: it keeps labeling rules in the hands of a single federal agency, which will decide what percentage of GMOs in a food product will require labeling; it allows for the use of obscure QR codes rather than clear labels on food packages; and most important, it kills far stricter rules written by states like Vermont, Connecticut, and Maine.

The Obama administration's fraught decision notwithstanding, the labeling debate continues to raise deeper questions about the ways our food is made. Do you really care only that a food was genetically engineered? Or would you also like to know that it was sprayed with an herbicide that is known to be carcinogenic to humans, or with another chemical known to destroy the plants that monarch butterflies need to survive? That it was sprayed with an insecticide known to kill bees? That it was grown in a monoculture field that is destroying biodiversity generally, or is polluting drinking water supplies? How far do you want to go with this?

When it comes to food labels, everything comes down to your level of risk tolerance, Jim Carrington, the president of the Danforth Center and a forceful proponent of the safety and benefits of genetic engineering, told me. Table salt is dangerous if used too much, and every year some people die from drinking too much water. Celery, broccoli, potatoes—lots of plants contain natural toxins that help them survive. Does that mean they deserve labels?

"The question is not whether something has the potential to cause cancer. There is nothing that is *not* in that category," Carrington said. "A rooster crows every morning and then the sun comes up. Association does not equal causation."

Carrington's view is that food production depends on all kinds of processes and ingredients that can be delivered in ways that are better or worse, and GMOs are no different.

"So let's say we label something that has a GMO ingredient," Carrington said, a note of sarcasm creeping into his voice. "If we require that, you know what I want to require? I want to know every input that went into that product. I'm concerned about water, and soil erosion, and nitrogen leaching into the waterways. That's all big-time environmentalism. Show me a label for everything in that box. Show me how much water the crops required, how much fertilizer ran into the nearest waterway or aquifer. But don't stop there. I want to know how many gallons of fuel were used per pound of produce, what the miles per gallon were for that tractor, whether or not there were any farm animals within two miles because I want to know about *E. coli*.

"Marking GMO ingredients as 'different' is marking something that in fact has no impact on what's in the box," he continued. "There is no substantive difference that will affect you. What I'm saying is, if you get to label something that has no bearing on your health or safety, I say let's go all the way. Show me every bit of information about how that product was produced so I, as a consumer, can make an informed choice. If you force a label on something that doesn't matter for reasons you say *do* matter—'I want to protect my children'—then I want to claim every bit of every other thing I'm concerned about. It's not rational, it's arbitrary, and it has negative consequences."

In a way, Carrington's modest proposal—labeling *everything* that goes into making our food—precisely reflects the sentiments of people who completely disagree with him about GMOs. It may be that our desire for labels is simply shorthand for our collective desire to know more about a food system that—to most of us—has become utterly industrial, technological, and abstract. We are given so little information about the way our food is grown and have so little contact with people or places that actually grow it. Perhaps the entire debate about GMOs may just be evidence of our cumulative ignorance about one of the most intimate things in our lives: the way we eat.

So how did we lose our way?

2.

The Long, Paved Road to Industrial Food, and the Disappearance of the American Farmer

The road we have traveled to our current state of eating is actually a very long, interconnected highway. After World War II, American national security strategists decided that protecting the homeland required building a network of broad interstates that mirrored the German Autobahn. This monumental road-building project—now close to 47,000 miles long—was initially conceived as a way to efficiently move troops and military machinery, but it has also had dramatic peacetime consequences for the American landscape, and for the American diet.

Suddenly, big, safe interstates—and the millions of miles of ring roads, state roads, and town roads they encouraged—allowed people to live farther and farther from the cities where they worked. People moved out of cities in droves, looking for new places to live. Land prices outside cities skyrocketed, and small farmers occupying that land had a hard time resisting when real estate developers came to call.

Suburban development hit small American farms like a virus. In the 1950s alone, some 10 million people left family farms. Chances are, your grandparents (or even your parents) can tell you stories about all those farms in your area that over the last few decades have been turned into subdivisions and shopping malls. In Maryland, where I live, suburban development has replaced 900,000 acres of farmland (and 500,000 acres of forest) in just the last forty years.

All these new roads, and the suburbs and industries to which they gave birth, caused a second tectonic shift in American culture: in the way we came to eat. Car-friendly fast-food chains like McDonald's and Carl's Jr. and Burger King started popping up along the new highways like weeds. By the early 1960s, Kentucky Fried Chicken was the largest restaurant chain in the United States.

These restaurants did not cook, exactly; what they did was heat up highly processed, prepackaged foods that tasted exactly the same, whether you were in Dallas or Des Moines. The ingredients didn't need to be fresh, they needed to be uniform, and storable, and—most important, given skyrocketing demand—they needed to be provided in vast quantities.

Fast-food joints didn't need local asparagus from New Jersey or collard greens from Georgia or one-of-a-kind apples grown in small orchards in New York. They needed commodity grains to sweeten their sodas, fry their fries, and feed the animals that could be turned into hamburgers and hot dogs and fried chicken. What these restaurants needed was corn, and wheat, and soybeans. And lots of them.

As small family farms near population centers went bankrupt or sold their land to developers, and as the American diet started demanding processed meals, food production flowed like beads of mercury to the control of larger and larger industrial farm operations in the Midwest. As food production became centralized, companies that controlled the grains, chemicals, and processing factories became bigger and much more politically powerful. Thanks to intensive lobbying, tens of billions of dollars in federal farm subsidies began flowing to giant agribusinesses that were driving the development of the industrial food system. As early as the 1970s, farmers around the country were being told (in the words of President Nixon's Agriculture Secretary Rusty Butz) to "get big or get out."

Most farmers got out. A little over a hundred years ago, there were 38 million people living in the United States, and 50 percent of them worked on a farm. Today, we have 300 million people. How many work on farms? Two percent.

Today, if you drive across the grain belt—Pennsylvania, Ohio, Indiana, Illinois, Iowa, Nebraska, Missouri, Kansas—you will spend many, many hours crossing an ocean of just three crops: corn, wheat, and soybeans. They are being grown by farmers you will likely never meet, processed in factories you will likely never see, into packaged foods containing ingredients that look nothing like the crops from which they were made. You won't see it, but your soda will be sweetened with high-fructose corn syrup, which replaced sugar in the 1980s. Your fries will be dunked in boiling soybean oil. And your burgers and nuggets and sliced turkey breast will all be processed from animals fed corn or soybeans, or both.

What you most likely won't see, out along on the great American road system, are regional food specialties, or the mom-and-pop diners and restaurants that used to serve them. New England clam chowder, New Orleans gumbo, Maryland crab bisque: all these foods require local ingredients, which (by definition) giant farms in Iowa or Kansas are unable to provide. Replacing them has been the food that these farms can provide: Fast food. Processed food. Soda. Pizza. Chicken nuggets. Cheap hamburgers. A vast culinary sameness, all essentially built out of two or three crops, controlled by a small handful of companies. All available twenty-four hours a day in any restaurant, dining hall, or gas station in the country.

It wasn't just fast-food restaurants pushing this new food system.

Food-processing giants like ADM, ConAgra, and Cargill learned to take monoculture corn and soybeans and turn them into the raw ingredients that could be made into just about anything a supermarket shopper wanted. Companies like General Mills or Coca-Cola could take a few cents' worth of wheat or corn and process it into Cocoa Puffs or a two-liter bottle of soda and sell it for a few dollars. As food scientists became more creative, they learned how to take wheat and corn and soy and turn them (along with the secret "fragrances" and "flavors" whose provenance only the food scientists seem to know) into limitless quantities of foods sold in suburban supermarkets—as often as not built on top of former farms.

These new foods were cheap to make, enormously profitable, and consumers seemed to love them. Americans spent $6 billion a year on fast food in 1970. By 2014, they were spending more than $117 billion. Today, Americans drink about 56 gallons of soda a year—about 600 cans per person—and every month, 90 percent of American children visit a McDonald's.

As industrial farms continued to grow, they gobbled up not just good land but marginal land, changing the face of millions upon millions of acres of forest, grasslands, hillsides, even wetlands. The strange thing was that the plants they grew—corn, soy, wheat—didn't seem to mind this change. The plants could grow, weed-like, even in marginal soil.

So, for better or worse, could the animals. Industrial feedlots across the Midwest began buying trainloads of corn and soybeans to feed an industry that now slaughters 9 billion animals a year.

As farms consolidated and grew, and as industrial processors increased their demand for ingredients that could be turned into shelf-stable food, farmers responded by growing what the market demanded—and eliminating what the market did not. Over the course of the twentieth century, the varieties of fruits and vegetables being sold by commercial U.S. seed houses dropped by 97 percent. Varieties of cabbage dropped from 544 to 28; carrots from 287 to 21; cauliflower from 158 to 9; tomatoes from 408 to 79; garden peas from 408 to 25. Of more than 7,000 varieties of apples, more than 6,200 have been lost.

THE DEVELOPMENT of American highways and suburbs caused one of the most dramatic changes in land use in the history of the world. But running parallel to this was an equally momentous shift in agricultural technology, which grew up fast to supply the rapidly changing American diet. In the 1930s, a plant breeder named Henry A. Wallace began boasting of the benefits of crossbred or "hybrid" corn, which he had meticulously developed to produce unprecedented yields. Even Wallace knew he was on to something dramatic. "We hear a great deal these days about atomic energy," he said. "Yet I am convinced that historians will rank the harnessing of hybrid power as equally significant."

Wallace was right. Corn yields doubled—from around 25 bushels per acre to 50 bushels per acre—in ten years. From 1934 to 1944—even before the postwar boom in agribusiness—hybrid corn seed sales jumped from near zero to more than $70 million, and rapidly replaced the enormous variety of seeds farmers had saved and traded for generations. By 1969, yields were up to 80 bushels an acre, and fully 71 percent of the corn grown in the United States was being grown from just a half-dozen types of hybrid seed. Industrial monoculture had arrived. Wallace's Hi-Bred Corn Company became Pioneer Hi-Bred International, America's largest seed company.

Since the 1960s, corn yields have doubled again, and now stand, in some places, close to 200 bushels per acre—nearly a tenfold increase in a single century. This phenomenal increase in production was dramatically accelerated by the invention, in the early twentieth century, of the Haber-Bosch process, which won its German inventors Nobel Prizes for discovering how to convert atmospheric nitrogen into ammonia. The ability to synthesize ammonia—routinely called the most important invention of the twentieth century—made it possible for industry to mass-produce two things that changed the world: explosives during the war and synthetic fertilizers after the war.

By the late 1940s, the war over, American industries found themselves with an enormous surplus of ammonium nitrate, the primary ingredient used to make TNT and other explosives. Since the synthetic compound also proved to be an excellent source of nitrates for plants, the U.S. Department of Agriculture (USDA) started encouraging the use of these chemicals on American farmland.

Suddenly, farmers (and their crops) shifted from a reliance on energy from the sun (in the form of nitrogen-fixing legumes or plant-based manure) to a reliance on energy from fossil fuels. Liberated from the old biological constraints, farms "could now be managed on industrial principles, as a factory transforming inputs of raw material—chemical fertilizer—into outputs of corn," Michael Pollan writes in *The Omnivore's Dilemma*. "Fixing nitrogen allowed the food chain to turn from the logic of biology and embrace the logic of industry. Instead of eating exclusively from the sun, humanity now began to sip petroleum."

A similar pattern emerged for the poison gases that industry had developed for the war: they were repurposed as agricultural pesticides and herbicides. Monsanto had begun the twentieth century making things like aspirin. In 1945, the company began making herbicides like 2,4-D, which would become a prime ingredient in Agent Orange, and is now one of the most popular farm sprays in the world. Monsanto also spent decades making PCBs, a compound used in both pesticides and electrical transformers (and long since banned as a dangerous carcinogen). By the 1960s, Monsanto was making a whole host of pesticides, with tough-sounding cowboy names like Lasso, Lariat, and Bullet. But the company's star product was Roundup, the glyphosate that is now the most popular herbicide in the world—and which, in a few short years, would be the star player in the growth of GMOs.

DuPont, Dow, Syngenta, Bayer, BASF—all the world's largest chemical companies made fortunes manufacturing compounds like DDT, atrazine, and scores of other farm chemicals. Today, the six top chemical companies control nearly 75 percent of the world's pesticide market.

This transition, from wartime chemicals to petroleum-based farm chemicals that now cover hundreds of millions of acres in the United States alone, has proven a double-edged sword for the world's farmers, and for the rest of us. For one thing, it means that most of us, in the words of the Indian food activist Vandana Shiva, are "still eating the leftovers of World War II."

True, it cranked up the amount of food farmers could grow, but it also (in the words of Czech-Canadian scientist Vaclav Smil) "detonated the population explosion." Farmers could now grow a lot more food, but suddenly—thanks in no small part to all this extra food—there were a lot more people to feed. Since the end of World War II, chemical fertilizer production jumped from 17 million tons per year to more than 200 million tons. Excess fertilizers and pesticides that are not taken up by plants seep into the rivers and bays, where they contaminate drinking water and cause algae blooms (and aquatic dead zones) so large they can be seen from space. They evaporate into the air, where they serve as major contributors to climate change.

And it's not just plants that these chemicals fertilize. Since their

Why?

advent, the human population has nearly tripled. Without these chemicals, Smil writes, billions of people would never have been born. The dousing of our crops with fossil fuels, in other words, meant we could now make unprecedented amounts of food. But now we had to.

For the large chemical companies, the global demand for more food provided a huge new market opportunity, not only for fertilizers and pesticides but for novel seeds to grow the crops themselves. By the late 1980s and early 1990s, the explosion of biotechnology—and especially in the ability of scientists to genetically engineer plants—meant that companies once devoted to chemistry began frantically shifting their emphasis to molecular biology. Chemical giants like Monsanto, DuPont, Syngenta, and Bayer began a frenzy of mergers and acquisitions, racing each other to dominate the world's seed industry. Monsanto's CEO Robert Shapiro moved especially aggressively in the mid-1990s, spending billions of dollars buying up seed companies and instantly making Monsanto the world's biggest ag-biotech company. The company bought Calgene, the maker of the Flavr Savr tomato, mainly because the smaller firm had ideas about GM cotton and canola.

Similar changes were under way at Dow and DuPont, which started out as makers of explosives like phenol and dynamite and are now two of the biggest GM seed companies in the world. In 1999, DuPont spent $7.7 billion to buy Pioneer Hi-Bred, which controlled 42 percent of the U.S. market for hybrid corn and 16 percent of the country's soybeans. The deal gave DuPont control of the world's biggest proprietary seed bank, as well as a global seed sales force.

The consolidation of the agrochemical giants has continued. In late 2015, DuPont and Dow Chemical announced a $130 billion merger, and Monsanto made a $45 billion offer to buy Syngenta.

The deal fell through, but Syngenta was immediately targeted by China National Chemical Corp., and Monsanto turned its attention to acquiring the crop science divisions of German chemical giants BASF and Bayer. Bayer responded in the spring of 2016 by offering to buy Monsanto for $62 billion. Monsanto rejected the bid as too low, but the companies remain in negotiations.

As late as the 1990s, the United States had hundreds of different seed companies; now we have a half-dozen. The biotech industry owns at least 85 percent of the country's corn seed, more than half of it owned by Monsanto alone. "This is an important moment in human history," Monsanto's CEO Robert Shapiro said in 1999. "The application of contemporary biological knowledge to issues like food and nutrition and human health has to occur. It has to occur for the same reason that things have occurred for the past ten millennia. People want to live better, and they will use the tools they have to do it. Biology is the best tool we have."

This, then, was the monumental shift that gave us GMOs. In a few short years, companies that had long known the power of chemistry discovered the power of biology. And the way we eat has never been the same.

GENETIC ENGINEERS are correct when they say that the fruits and vegetables we see in the supermarket look nothing like their wild forebears. The tomatoes we eat today—juicy and sweet, not bitter and toxic—are the result of thousands of years of human selection. So is the corn. The first cultivated carrots—typically yellow or purple—were grown in Afghanistan. It was only after traders carried them to Europe and the Mediterranean, where they were crossed with wild varieties, that their offspring gradually turned orange.

In the nineteenth century, the Austrian monk and scientist Gregor Mendel discovered how a plant passed its traits from parent to offspring. Taking anthers from one variety and dusting them with pollen from another, he crossed some 10,000 plants: round peas with wrinkled peas; peas from yellow pods with peas from green pods; peas from tall and short plants. Every trait a plant's offspring exhibited—height, color, shape—depended on what Mendel called "factors" that were either dominant or recessive. So if a round pod was crossed with a wrinkly pod, three out of four times the offspring would be round, meaning that was the dominant trait. The last one could either be round or wrinkly. That's because these factors apparently came in pairs, one from each parent, and were inherited as distinct characteristics.

DNA was known to be a cellular component by the late nineteenth century, but Mendel and other early geneticists did their work without understanding its role in heredity. By the late 1940s, most biologists believed one specific kind of molecule held the key to inheritance, and turned their focus to chromosomes, which were already known to carry genes. As agricultural research began moving from the field into the laboratory, scientists discovered a new way to mirror natural selection: by exposing plants to chemicals or radiation, they could alter the plant's biochemical development. They could force it to mutate. By some estimates, radiation mutagenesis has introduced some 2,500 new varieties of plants into the world, including many that find their way onto our plates, like wheat, grapefruit, even lettuce.

With the flowering of genetic engineering in the 1970s and 1980s, scientists figured out how to go into an organism—a plant or an animal, a bacteria or a virus—remove one or more genes, and stitch them into the genetic sequence of another organism. This process became known as recombinant DNA technology.

The first commercially available product of genetic engineering was synthetic insulin. In humans, insulin is normally made by the pancreas and helps regulate blood glucose; produce too little insulin, and you can develop type 1 diabetes. Traditionally, increasing a diabetic's insulin required collecting insulin from the pancreatic glands of pigs or cattle, a problem not only for the animals but also for people who became allergic to the insulin's different chemical structure.

In 1978, scientists at the company Genentech used genetic coding to create a synthetic insulin known as humulin, which hit the market in 1982. Today, this GM insulin is produced around the clock in giant fermentation vats and is used every day by more than 4 million people. Similar technology has been used to produce vaccines that combat hepatitis B; human growth hormone, which combats dwarfism; and erythropoietin (EPO), which helps the body produce red blood cells (and has been, illegally, used to boost racing performance by riders in the Tour de France).

In the late 1980s, genetic engineers turned their sights on cheese. Just a few years before the release of the Flavr Savr tomato, the combination of a single gene from a cow was stitched into the genome of a bacterium (or a yeast) to create rennin, a critical enzyme in the production of hard cheeses. Once obtained as a by-product of the veal industry, rennin was traditionally collected from the lining of a cow's fourth stomach. GM rennin is now used in some 90 percent of the cheese made in the United States.

But compared with what was to come, these early experiments were, well, small potatoes. The real money, agrochemical companies knew, would come through genetically engineering the crops Americans ate most. Not cheese, but corn and soybeans. Control those crops, and you could dominate a fundamental part of the global economy.

Monsanto's most important push was to create seeds the company could sell alongside Roundup, already the bestselling farm chemical in the world. Creating (and patenting) Roundup-resistant seeds would secure the company's global share in seeds *and* herbicides. The world's farmers wouldn't buy just one. They would buy both.

"It was like the Manhattan Project, the antithesis of how a scientist usually works," said Henry Klee, a member of Monsanto's Roundup research team. "A scientist does an experiment, evaluates it, makes a conclusion, and goes on to the next variable. With Roundup resistance, we were trying twenty variables at the same time: different mutants, different promoters, multiple plant species. We were trying everything at once."

It took four years, and a bizarre eureka moment, for Roundup Ready seeds to be born. Frustrated in their lab work, company engineers decided to examine a garbage dump 450 miles south of Monsanto's St. Louis headquarters. There, at the company's Luling plant on the banks of the Mississippi, the engineers found plants that had somehow survived in soil and ponds near contamination pools, where the company treated millions of tons of glyphosate every year. The hardiest weeds were collected, their molecular structure examined, their genes replicated and inserted into potential food crops.

When Roundup Ready soybeans were finally launched, in 1996, they instantly became an essential part of a $15 billion soybean industry. Roundup Ready soybeans covered 1 million acres in the United States in 1996; 9 million acres in 1997; and 25 million in 1998. Today, 90 percent of the country's 85 million acres of soybeans are glyphosate resistant.

The first insecticide-producing corn plant was approved in 1996, the same year Monsanto released its Roundup Ready soybean.

Today, the overwhelming majority of the GM crops grown in the United States—some 170 million acres of them—are still grown to feed the industrial food system. In Iowa, GM corn is grown to feed the numberless cows and pigs that enter into the fast-food system. In Maryland, GM soybeans are grown to feed the hundreds of millions of chickens on the state's Eastern Shore, which will enter the same system. In Nebraska, GM canola is grown to make the oil to fry the french fries served in the country's galaxy of drive-through restaurants.

Why are the crops genetically engineered? For the same reason the highways were built: they make everything faster, more uniform, more efficient. In the United States, GM crops are grown mainly for two reasons: to increase yields and—especially—to allow farmers to spray their crops with chemicals that kill insects, diseases, or weeds. By developing crops that can withstand regular pesticide dousing (or, like Bt corn, that can provide their own insecticide), scientists have enabled farmers to eliminate everything but the crops whose numbers they are trying to maximize. Gone are the weeds. Gone are the insects. The whole system works—in the most literal sense—like a well-oiled machine.

Food and chemical companies—and the farmers who grow for them—say that GM crops allow them to deliver a lot of food to a lot of people for very little money, and this is true, as far as it goes. Americans have become very comfortable spending relatively little money for their food. According to the World Bank, Americans spend considerably less per capita on food than anyone else in the world. Food expenses are much higher in the UK (9 percent), France (14 percent), South Africa (20 percent), and Brazil (25 percent). And our food is cheap not just compared with other countries; it's cheap compared with the food we used to eat, before all our small farms moved to the Midwest. In 1963, the year I was born, Americans were spending close to a third of their income on food. Now we spend about 6 percent.

SO HERE WE ARE. Genetic engineering did not create any of the structures that hold up our current food system. It merely added a set of tools—very powerful tools—to keep the whole machine running. The fact that these tools arrived on the scene at the very moment that the American food economy was becoming so intensely industrialized has created both enormous profits for the companies and enormous health and environmental problems for the rest of us. Had genetic engineering come about at a different time—were we still a nation of small farmers, for example, and were biotech companies making seeds to help local farmers grow nutritious produce—things might have turned out entirely differently.

But that's not what happened. When it comes to GMOs, it's impossible to separate science from industry, or industry from politics. It's all tangled up together, and we are eating all of it. The argument that genetic engineering is just another step in a tradition of plant breeding that goes back 10,000 years is absolutely true. But it is also true that biotechnology has developed at a time when its primary use has been to fuel a food system that is far bigger, more complex, and more destructive than anything the world has ever seen.

Because this system has become so profitable, companies have gone to great lengths to cement their control over it in all three branches of the federal government. Through the White House, they push their own people to the top of federal regulatory agencies. In Congress, they use lobbyists and political muscle to influence policy, and to keep federal farm subsidies flowing. In the courts, beginning in 1980, they have repeatedly convinced judges that they deserve patents (to quote a famous court decision) on "anything under the sun that is made by man." To date, tens of thousands of gene patents have been awarded to biotech companies, and tens of thousands more wait in the wings. This means, in the most fundamental way, that our food supply is owned and controlled by a very small handful of companies.

This is nowhere more evident than in the hundreds of billions of taxpayer dollars that move through federal regulatory agencies into the hands of companies these same agencies are supposed to regulate. Between 1995 and 2010, large agricultural companies received $262 billion in federal subsidies, a great percentage of it going to companies developing GM food products.

It is also evident in the way federal agencies view their relationship with the companies they are charged with overseeing. Since the 1980s, regulation of GMOs has been handled through a complex web of three vast federal agencies. A genetic engineer has to get a permit from the USDA to field-test a GM crop. Then—after several years of trials—the engineer must petition for the deregulation of the crop. If the crop has been designed to be pest-resistant, the EPA will regulate it as pesticide and demand more data. Finally, the FDA evaluates the plant to make sure it is safe for consumption by people or animals.

But in reality, safety testing of GMOs in the United States is left to the companies that make them. This is very much in line with much of American regulatory policy and is dramatically different from the approach taken in Europe, where regulators require that the introduction of GM foods should be delayed until the long-term ecological and health consequences of the plants are better understood. In the United States, industry and government have decided that GMOs are "substantially equivalent" to traditional foods, and therefore should not be subjected to new federal oversight.

U.S. policy "tends to minimize the existence of *any* risks associated with GM products, and directs the agencies to refrain from hypothesizing about or affirmatively searching for safety or environmental concerns," legal scholar Emily Marden writes.

Federal Government: Watchdog or Cheerleader?

The shift in federal policy from "regulating" GMO foods to "promoting" them was subtle, and to most of the country, entirely invisible. Back at the beginning, in 1974, Paul Berg, often called the father of genetic engineering, persuaded other molecular biologists to be cautious in the pioneering work they were doing in their laboratories. "There is serious concern that some of these artificial recombinant DNA molecules could prove biologically hazardous," Berg wrote at the time. To address these questions, Berg and his colleagues at the National Academy of Sciences urged caution in the development of genetic engineering technology until scientists could form standards for biological and environmental safety. Addressing the technology itself, rather than its application to food production, the now famous "Berg Letter" acknowledged that such a cautious approach was based on "potential rather than demonstrated risk," and might well mean the "postponement or possible abandonment" of some ongoing experiments.

"Our concern for the possible unfortunate consequences of indiscriminate application of these techniques," Berg wrote, "motivates us to urge all scientists working in this area to join us in agreeing not to initiate experiments until attempts have been made to evaluate the hazards and some resolution of the outstanding questions has been achieved."

After Berg's letter was published, a group of scientists organized a closed-door conference at Asilomar, California, in February 1975 to formulate research guidelines that would prevent health or ecological trouble from rippling out from this new technology. But the letter also made it very clear that scientists themselves, and not the government, would be in charge of keeping an eye on things. No new legislation was needed, the letter noted. Scientists could "govern themselves."

James Watson, one of the discoverers of the double helix structure of DNA and an attendee at the Asilomar conference, made it clear that scientists were not interested in ethical guidance from outside the profession. Although some "fringe" groups might consider genetic engineering a matter for public debate, the molecular biology establishment never intended to ask for guidance. "We did not want our experiments to be blocked by over-confident lawyers, much less by self-appointed bioethicists with no inherent knowledge of, or interest in, our work," Watson wrote. "Their decisions could only be arbitrary."

Watson had nothing but contempt for those who would stand in the way of scientific research; he once referred to critics of genetic engineering as "kooks, shits, and incompetents." The risks from this technology, he wrote, were about the same as "being licked by a dog."

The National Institutes of Health quickly adopted the Asilomar conclusions and turned them into a national research standard: biotechnology research would be largely self-regulated and should be encouraged, not hampered, by federal oversight.

At first, most of the research being done in biotechnology had to do with medical research, not food production, and given the lack of public debate on the issue, few health or environmental groups paid much attention to genetically engineered food. But within a few years, the potential applications—and the potential profits—in agriculture became obvious. The question was, what would happen once this technology escaped the laboratory and was scaled up to reach all our dinner tables?

"In the 1970s, we were all trying to keep the genie in the bottle," said Arnold Foudin, the deputy director of biotechnology permits at the USDA. "Then in the 1980s, there was a switch to wanting to let the genie out. And everybody was wondering, 'Will it be an evil genie?'"

The genie was released in the 1980s and 1990s by the Reagan and Bush administrations, which had long made industrial deregulation a national priority. To their eyes, the burgeoning biotech industry was a perfect merging of business and science that—if left alone—would generate colossal corporate profits for American agricultural conglomerates.

"As genetic engineering became seen as a promising investment prospect, a turn from traditional scientific norms and practices toward a corporate standard took place," sociologist Susan Wright observes. "The dawn of synthetic biology coincided with the emergence of a new ethos, one radically shaped by commerce."

If nothing else, all this grain would supercharge the meat industry: Reagan's first secretary of agriculture was in the hog business; his second was president of the American Meat Institute. George H. W. Bush later appointed the president of the National Cattlemen's Association to a senior USDA position.

The trick was to come up with federal policy that would allow this new technology, and the products it generated, to enter the marketplace without regulatory hassles—and without worrying the public that the foods they made were somehow different from traditional foods.

Creating these rules required some fancy bureaucratic footwork. Since 1958, Congress (through the Federal Food, Drug, and Cosmetic Act) had mandated that "food additives"—typically chemical ingredients added to processed foods—should undergo extensive premarket safety testing, including long-term animal studies. Commonly used ingredients, like salt and pepper, were considered GRAS (for "generally regarded as safe") and were exempted from further testing.

The billion-dollar question was: Should genetically altered foods be considered a "new" food additive—and thus be forced to undergo extensive testing—or "safe," like salt and pepper?

In the early 1990s, the FDA put together a scientific task force to study this question. A consensus quickly emerged that these new products should be developed cautiously, and should be tested to see just what impact they might have on the health of people and animals who eat them.

"The unintended effects cannot be written off so easily by just implying that they too occur in traditional breeding," wrote microbiologist Dr. Louis Pribyl. "There is a profound difference between the types of unexpected effects from traditional breeding and genetic engineering."

Pribyl said applying the GRAS label to GMOs was not scientifically sound. It was, instead, "industry's pet idea"—a way to apply a formal stamp of government approval on foods that were, in fact, a completely new thing under the sun.

The director of the FDA's Center for Veterinary Medicine went further, warning that using GMOs in animal feed could introduce unexpected toxins into meat and milk products. The head of the FDA's Biological and Organic Chemistry Section emphasized that just because GMOs had not been proven to be dangerous did not confirm their safety. Saying that GMOs were as safe as traditional foods "conveys the impression that the public need not know when it is being exposed to new food additives."

Likewise, deep inside the labs of government and university laboratories, enthusiasm for genetically engineered food was not nearly as uniform as its promoters in government or industry claimed. "This technology is being promoted, in the face of concerns by respectable scientists and in the face of data to the contrary, by the very agencies which are supposed to be protecting human health and the environment," said Suzanne Wuerthele, a toxicologist at the EPA. "The bottom line in my view is that we are confronted with the most powerful technology the world has ever known, and it is being rapidly deployed with almost no thought whatsoever to its consequences."

But given the revving engines of industry, it was tough for GMO skeptics in the scientific community to have their voices heard. University scientists applying for grants to look more closely at potential dangers of GMOs were routinely underfunded, squashed, or simply shouted down. Government scientists were stymied by the influence of the food and chemical industries, whose former executives were routinely placed at the top of the very agencies charged with regulating products made by the companies they used to work for.

It was no secret that the Reagan and Bush administrations had made subsidizing (and deregulating) these companies, and this technology, a national priority. There was no way a regulatory agency could fairly scrutinize an industry it was also funding with so much money, said Philip Regal, a professor at the University of Minnesota's College of Biological Sciences.

"The more I interacted with biotech developers over the years, the more evident it became that they were not creating a science-based system for assessing and managing risks," Regal said. "And as momentum built and pressures to be on the bandwagon mounted, people in industry and government who were alerted to potential problems were increasingly reluctant to pass the information on to superiors or to deal with it themselves. Virtually no one wanted to appear as a spoiler or an obstruction to the development of biotechnology."

From biotechnology's earliest days, "it was clearer than ever that the careers of too many thousands of bright, respected, and well-connected people were at stake—and that too much investment needed to be recovered—for industry or government to turn back," Regal said. "The commercialization of GE foods would be allowed to advance without regard to the demands of science; and the supporting rhetoric would stay stretched well beyond the limits of fact."

Despite this backbeat of scientific concern, the administration of George H. W. Bush—with considerable input from policy executives at companies like Monsanto—ruled that GMOs would not be subjected to any more testing than traditional foods. The GRAS policy would be explicitly designed not to test new food science but to assure "the safe, speedy development of the U.S. biotechnology industry," Bush's FDA Commissioner David Kessler wrote.

The FDA policy made it official: the agency was "not aware of any information showing that foods derived by these new methods differ from other foods in any meaningful or uniform way, or that, as a class, foods developed by the new techniques present any different or greater safety concern than foods developed by traditional plant breeding."

Genetic manipulation was no different from breeding techniques farmers had been using for centuries, Kessler argued. Properly monitored, GM foods posed no special risk and should not require advance federal approval before being sold. "New products come to our kitchens and tables every day," Kessler said. "I see no reason right now to do anything special because of these foods."

Beyond giving a green light to the technology itself, the FDA even left the decision about whether new GM foods were GRAS to the companies themselves.

The victory of agribusiness over the FDA's own scientists furthered a decades-long tradition, in which government agencies "have done exactly what big agribusiness has asked them to do and told them to do," a fifteen-year veteran of the FDA told *The New York Times*. "What Monsanto wished for from Washington, Monsanto—and by extension, the biotechnology industry—got."

Genetic engineering now had the full-throated support of the U.S. government. Administration officials and food industry groups of all kinds lined up to tout the benefits of biotechnology and celebrate the wall that had been erected to protect companies from federal oversight. The hands-off approach was framed as a fine example of what Bush administration officials called "regulatory relief."

Such policy "will speed up and simplify the process of bringing better agricultural products, developed through biotech, to consumers, food processors, and farmers," Vice President Dan Quayle announced. "We will ensure that biotech products will receive the same oversight as other products, instead of being hampered by unnecessary regulation."

Quayle's declaration put the full weight of the federal government behind a policy that had largely been dictated by agribusiness—especially Monsanto, which by this time had become the world's largest developer of GM seeds.

"What Monsanto wanted (and demanded) from the FDA was a policy that projected the illusion that its foods were being responsibly regulated but that in reality imposed no regulatory requirement at all," writes Steven Druker, an environmental attorney and author of the book *Altered Genes, Twisted Truth*. The FDA "ushered these controversial products onto the market by evading the standards of science, deliberately breaking the law, and seriously misrepresenting the facts—and that the American people were being regularly (and unknowingly) subjected to novel foods that were abnormally risky in the eyes of the agency's own scientists."

In the twenty-five years since the GRAS decision, the FDA has never overturned a company's safety determination and, thus, has never required food-additive testing of any transgenic crop.

The Perils of Self-Regulation

Allowing industry to regulate itself has led to a great deal of criticism, of course, since it's rarely in industry's best interest to reveal problems with its products, even when they are well known. This is not a new game, as users of countless other products—from Agent Orange to cigarettes to opioid pain relievers—have learned. In those cases, industry scientists knew their products were harmful, but companies continued to promote them and withhold conflicting data for years.

In 2002, a committee of the National Academy of Sciences, the country's premier scientific advisory body, declared that the USDA's regulation of GMOs was "generally superficial": it lacked transparency, used too little external scientific and public review, and freely allowed companies to claim that their own science was "confidential business information." The committee itself complained that it was denied access to the very information it needed to conduct its review—and not just by the companies. The amount of information kept secret by the USDA itself "hampers external review and transparency of the decision-making process."

The EPA has also come under intense criticism for—among other things—the way it has handled the staggering population declines of bees and monarch butterflies, both of which have been linked to chemicals sprayed on hundreds of millions of acres of GM crops. The monarch is now as much a symbol for the anti-GMO movement as the polar bear is for climate change activists.

And the FDA? The agency's own policy states that "it is the responsibility of the producer of a new food to evaluate the safety of the food." Denied essential proprietary data by companies they are supposed to oversee, the FDA "is unable to identify unintentional mistakes, errors in data interpretation or intentional deception, making it impossible to conduct a thorough and critical review," a study by William Freese and David Schubert at the Center for Food Safety found.

Such voluntary self-regulation means "government approval" amounts to little more than a rubber stamp, Michael Hansen of the Consumers Union told me. Even though companies test their products, they have a way of doing tests over and over until they get the results they like, Hansen said, and show only favorable results to the agencies overseeing them. Companies are not always forthcoming with the data they do accumulate, and sometimes actively refuse to turn over research even when federal regulators ask for it.

For four decades, the American legal system has repeatedly upheld the industry's right to control the seeds underpinning our food. Monsanto alone filed 147 seed patent infringement lawsuits in the United States between 1997 and April 2010, settling all but nine out of court. The cases that went to court were all decided in favor of the company.

North of the border, where GMOs are considerably less popular, it has been a bit more complicated.

Patenting Our Food: The Schmeiser Case

Wandering around his canola farm in Saskatchewan in the late 1990s, Percy Schmeiser noticed plants growing not just in the fields, but in a nearby drainage ditch. He did what many farmers would have done to get rid of an unwanted infestation: he sprayed the plants with glyphosate.

Nothing happened.

The canola plants, it turned out, had sprouted from genetically modified Roundup Ready seeds that had floated in from nearby farms. Schmeiser's neighbors—and 30,000 other Canadian farmers—had paid Monsanto $15 an acre for the right to use these GM seeds; their harvests constituted more than 40 percent of Canada's canola crop. Monsanto was keen to protect its product: they had farmers sign contracts agreeing not to save or replant the seeds, and they sent out inspectors to make sure the farmers were complying with seed contracts.

Schmeiser had not been part of this deal. For several years, he had planted his own (non-GM) canola fields with seeds he had saved from his own plants. After discovering Roundup-resistant plants in his ditch, he wondered just how much of his farm had been contaminated by Monsanto's seeds. He sprayed three acres with glyphosate; 60 percent survived.

When word got out that Schmeiser had acres of Monsanto-patented seeds growing in his fields, someone called the company, using an anonymous-tip line the company had set up for farmers to turn in their neighbors. Monsanto sent private investigators to patrol the roads near Schmeiser's farm. They took crop samples from his fields, and in 1998, Monsanto notified Schmeiser that he was using the company's seeds without a license.

Undaunted, Schmeiser saved seeds he had harvested from plants that had survived spraying, and planted them on about 1,000 acres. Later tests would confirm that nearly 98 percent of these plants were Roundup resistant.

Monsanto sued Schmeiser for patent infringement. "We've put years, years, and years of research and time into developing this technology," said Randy Christenson, Monsanto's regional director in Western Canada. "So for us to be able to recoup our investment, we have to be able to pay for that."

Schmeiser had a different take. "I've been farming for fifty years, and all of a sudden I have this," he said. "It's very upsetting and nerve-racking to have a multi-giant corporation come after you. I don't have the resources to fight this."

In court, Schmeiser argued that the GM seeds on his field had arrived the same way seeds have always arrived—they were blown in on the wind. "You can't control it," he said. "You can't put a fence around it and say that's where it stops. It might end up 10 miles, 20 miles away." Furthermore, he argued, a company should not be allowed to patent a higher life form, like a canola plant. Plants were part of the natural order of things, not widgets that came off a company's factory floor.

This was not the first time Canada's courts had to wrestle with whether a company could own a life form. The country's supreme court had previously ruled that Harvard University did not have the right to patent a genetically altered "OncoMouse" (a rodent genetically designed to rapidly develop cancer) even though it had taken university scientists seventeen years to develop it. Courts in Europe and the United States had sided with Harvard, but Schmeiser argued that the Canadian ruling—that an advanced life form could not be patented—ought to apply in his case too.

In 2001, a trial judge rejected Schmeiser's argument and fined him $20,000 for infringing on Monsanto's patent. On appeal—and in a show of just how complex biology and patent law can be—the Supreme Court agreed, but its 5–4 decision was split. The court's minority argued that Monsanto had claimed patent protection over the *gene* and the *genetic process*, not the life form (that is, the plant itself), and that Schmeiser should not be held liable for using an (unpatentable) plant. The majority, interestingly, agreed: *plants* could not be patented in Canada. But the five justices also ruled that a plant's genes *could* be patented: by "using" the plant, Schmeiser had in effect "used" the patented gene.

The ruling forced Schmeiser to turn over any Roundup Ready seeds or crops on his property. In a small consolation, the court ruled that Schmeiser did not have to pay Monsanto for the profits he had made from his crop. Monsanto, for its part, made sure the world knew its point of view. "The truth is Percy Schmeiser is not a hero," the company says. "He's simply a patent infringer who knows how to tell a good story."

In effect, the Canadian court gave Monsanto legal control over something it could not patent—Roundup Ready canola plants—by giving it legal control over something it could control—the plant's genes. In Canada at least, plants themselves are still not patentable, and farmers are allowed some protection if they don't intentionally use patented seeds. Canadian growers also (for the moment, at least) still enjoy a "farmer's privilege"—protected by the national Plant Breeders Rights Act—that allows them to save and replant traditionally bred seeds.

This is in direct contrast to laws in the United States, where the

Supreme Court has ruled that plants can be patented *in spite* of laws protecting a farmer's right to save seeds. This position—promoting the rights of large companies over the rights of small farmers—is very much in keeping with the American government's longstanding and unwavering support of the biotech industry.

In the end, the early (and ongoing) rush to develop, plant, and profit from GM seeds has simply outpaced and overwhelmed our ability to understand their impact on our lives, John Vandermeer, the agroecologist at the University of Michigan, told me.

"I would be far less negative about GMOs had the people developing [them] taken the same approach as they did at Asilomar," Vandermeer told me. "They could have said, 'Let's have a moratorium on selling them until we can be sure that they are safe.' But partly because of the profits involved, that was never done. Had we done this, it's my guess that Bt and Roundup transgenic crops never would have been developed and spread throughout the landscape. What we are discovering is that they should probably never have been used."

It's not just that such company-directed testing might miss (or cover up) a dangerous product. Chemical-intensive farming has also led to a tremendous loss of biodiversity—both above and below ground, Vandermeer says. From the massive genetic biodiversity of traditional agroecosystems, we now have millions upon millions of acres planted with the same hybrid corn variety. Soil, a fantastic ecology of interdependent living organisms, has been reduced to a medium "as devoid of life as possible," Vandermeer writes.

This positive feedback loop—vast acreage planted with single crops, all propped up by rivers of chemical fertilizers, which then cause the monocultures to flourish—has also created a dramatic increase in the potential for collapse. All it takes is for an insect (or a virus) to pick the lock of a plant's defenses, and an entire crop can disappear. A nineteenth-century blight in Ireland ruined a potato crop, and fully a million people starved to death. In the 1950s, the Gros Michel banana—planted on monoculture plantations across Latin America—was virtually wiped out by a fungus. Today, the Cavendish—the Gros Michel's successor and likely the only banana you have ever eaten, whether you've eaten it in Los Angeles, New York, London, or Hong Kong—is grown on vast plantations in Asia, Australia, and Central America. And a fungus, called Tropical Race 4, has picked its lock. Unless breeders (including geneticists) can figure out a way to get banana trees to develop resistance, there will likely come a day very soon when we—outside the tropics, at least—will have no more bananas.

Here at home, this system also means that companies get to decide what products to create. In the United States, GMOs are designed more to make corn for cheese puffs and cheap hamburgers than to develop nutritionally dense food for people either here or in developing countries. Such uses cheapen the promise of food technology by using it to create empty calories and poor nutrition, serving industry profits but not the general welfare of either people or the planet.

Without broader research conducted outside the food industry itself, the editors of the scientific journal *Nature* say, the development of genetic engineering "will continue to be profit-driven, limiting the chance for many of the advances that were promised thirty years ago—such as feeding the planet's burgeoning population sustainably, reducing the environmental footprint of farming and delivering products that amaze and delight."

Leaving the power of GM technology to a group of global food conglomerates is plainly problematic for a whole array of reasons.

But there are small pockets out there, mostly in university and other nonprofit research labs, where an entirely different approach to genetic engineering is taking place. Because while most GMOs currently bolster the production of cheap, unhealthy, processed food, there are scientists at work developing foods that could actually change the world for the better.

Mapping and Engineering and Playing Prometheus

As you walk into the Delaware Biotechnology Institute, the first thing you see is a giant double helix engraved on a large piece of Plexiglas. Inside, adorning the walls, are vivid, Technicolor photographs that look like images beamed back from the Hubble Space Telescope: streaks and smears of purples, greens, and reds that could be gas clouds swirling through star clusters. They aren't. They are pictures of cellular components like mitochondria, taken with nanoscale bioimaging so impressive that its inventor won a Nobel Prize—and so sensitive that an entire wing of the building had to be built on a special slab to prevent vibrations from ruining photographic precision.

Deep inside the building, Blake Meyers leads me through a room given over to racks of whirring computers. There are wires and carcasses of old machines everywhere, and a power generator the size of three refrigerators, whose excess heat is balanced by an air-conditioning unit mounted on the building's roof.

These computers are "energy hogs," Meyers said; if the power goes out, the backup batteries can support them for only about fifteen minutes. The day I visited, Meyers said one of his servers had been crunching data for one project for six weeks straight.

Meyers is a prominent plant geneticist, with degrees from the University of Chicago and UC-Davis, who also did postdoc work for the agrochemical giant DuPont. He is a vegetarian and deeply conscious of environmental problems. Engineering new kinds of plants, he says, could fix problems on a global scale.

Take nitrogen fixation, Meyers said. Modern agriculture uses huge amounts of natural gas to make and add nitrogen fertilizer to fields growing corn, because corn plants suck so many nutrients from the soil. But if you could engineer corn to pull nitrogen straight out of the atmosphere, think of how much synthetic nitrogen you could stop making and applying.

"This is science fiction right now, but if we could produce corn that fixes nitrogen, we could eliminate hundreds of millions of units of natural gas, plus eliminate massive environmental sources of nitrogen that currently are emitted or run off," he said. "You would have done a really beneficial thing. You would be one huge step closer to growing corn under organic conditions."

Meyers lists other promising projects: creating a calorie-dense rice that also fixes its own nitrogen *and* resists insects *and* resists drought. "Think of what you could do," he said. "You could create a supercrop. But you're not going to get there through natural selection and traditional breeding. This would be like super-speeding evolution."

Meyers is also entirely skeptical about the ability of small-scale farming to feed the world.

"We're looking at a future with 9 billion people, and it may be as many as 12 billion—how are we going to feed them, plus generate sustainable fuels and bioproducts?" Meyers asked me. "Through small-scale farming? There's not an ice cube's chance in hell that we can do that. We need substantial bumps in agricultural productivity if we're going to provide the resources that people are expecting. Addressing that need is going to take every tool in our toolbox. That's why we think GM technology has to play a role in this."

Inside his laboratory, Meyers drew me near a machine and asked a lab assistant to punch up a screen. There appeared before me an image of a grid, three squares by three, labeled "Tile 13." Again, if I hadn't known better, I might have assumed I was looking at a photograph of distant galaxies: blurring patches comprising thousands of tiny spots of black, white, and gray.

What we were looking at was in fact a constellation of "small RNAs" from soybean leaves. It is in this laboratory that Meyers—like scientists all over the world—is either (depending on your point of view) extending a long tradition of plant breeding or taking a Promethean leap once left to the gods.

The Genetic Equivalent of War and Peace

Understanding genetics is frequently compared to learning a language. Just as letters and words are arranged in certain patterns to transmit information on the page (the metaphor goes), so are microscopic elements inside cells arranged in predictable ways. Once these elements have been assigned letters, their combinations (or sequences) can be read just like words, sentences, or entire volumes. And just as the twenty-six letters in the English language can lead to more than a million words (and a virtually infinite number of unique sentences, paragraphs, and books), so do the four letters at the foundation of genetics lead to an unimaginably diverse number of organisms.

The "letters" of genetics, called nucleotides, are A (adenine), T (thymine), C (cytosine), and G (guanine). Inside the nucleus of all living cells, these nucleotides form chains that microbiologists can read: AATTCCGG, for example. Given their chemical makeup, each individual nucleotide is attracted to a very particular mate: A (which has two rings of nitrogen) will bond only with T (a single ring); the two-ring G will pair only with the single-ring C. This makes the pairing of letter chains very predictable: our chain of AATTCCGG, for example, will bond (and form a genetic "word") with a complementary, mirrored chain of TTAAGGCC.

Complete sequences (the "chapters") of these genetic words are called chromosomes, which can be made up of thousands of different genes (and therefore millions of individual nucleotides). Chromosomes are so densely packed that, linked together and stretched out, the DNA molecules in just one of your cells would be taller than you are. Lined up end to end, all the DNA in all your cells would stretch—in a very thin line—some 6 billion miles.

All together, an organism's chromosomal chapters make up its entire book-length genome. It took about 3.1 million letters, 588,000 words, and 365 chapters to make *War and Peace*. It takes 3.2 billion nucleotide base pairs, 19,000 genes, and 23 pairs of chromosomes to make a human genome. More relevant for Blake Meyers is the soybean genome, which has more than 1.1 billion base pairs, 46,000 genes, and 20 chromosomes. Or the maize genome, which has 2.3 billion base pairs and 32,000 genes.

Counting genes is one thing. Understanding how they work has been another thing entirely. Scientists once summed up the way genes control cell function with a simple formula known as the Central Dogma: DNA codes for RNA and RNA codes for protein. DNA contains a cell's blueprint; RNA transmits the blueprint to create proteins; proteins carry out a cell's functional tasks, which in turn determine the structure and behavior of the organism itself.

As our understanding of genetics has sped up, however, new molecular worlds have opened up. When scientists sequenced the human genome a decade ago, it was somewhat like looking at a blueprint in a foreign language—everything was marked in its proper location, but no one could tell what it all meant. Less than 2 percent of our genome seemed to code for proteins that actually do anything, so the vast majority of our DNA has been like biology's dark matter, acting in ways that remain mysterious and only partially understood. For years, long stretches of noncoding genes were simply tossed off as "junk DNA."

That view has changed. A five-year project called ENCODE— for "Encyclopedia of DNA Elements"—found that as much as 80 percent of the human genome is biologically "functional," meaning that even if certain genes don't directly code proteins, they can still influence how nearby genes are expressed, and in which types of cells. These noncoding regions of DNA can have major bearing on diseases and genetic mutations. Because the genes of an organism are interconnected, a single disturbance in gene organization (or function) can affect multiple gene systems. This has potentially serious implications for cellular function and the overall health of the organism. Consider that altering a single letter of the genetic code of a single gene can be a significant step leading to cancer—a disease that involves alterations in the function of multiple genes, proteins, and cellular systems.

So if every cell in an organism has the same DNA—if the cells in your eyes contain the genes for your toes, and vice versa—why is it that cells are so different from one another, and do one thing, and not another? In other words, why do cells allow you to see out of your eyes and not out of your toes? In the plant world, why is a leaf cell not the same as a root cell or a flower cell? Understanding this requires going back to the idea of genetics as both a code (an "alphabet") and a language (a mode of "expression").

The answer lies in the way genes are expressed. Since every cell in an organism contains exactly the same genes, it is in this "expression"—as genes are either "turned on" or "turned off"—that cells become distinct. It's gene expression that causes them to become either a root cell or a leaf cell, and collectively create plants—and whole other organisms—that are distinct. Changing a single letter can make a huge difference. Just as the difference between the words "tasty" and "nasty" is a single letter, so (in humans) a slight shift in nucleotide sequences could cause changes in amino acids, the building blocks of proteins, that can cause sickle-cell anemia, or Parkinson's, or Alzheimer's.

The code itself, the sequence of A's, C's, G's, and T's, is inscribed in DNA. But for this code, known as an organism's genotype, to orchestrate an organism's structure and behavior, known as its phenotype, the code must first be "transcribed" (inside the nucleus) from DNA to RNA and then "translated" (outside the nucleus) into a protein. And that requires an understanding not just of DNA, but of RNA as well.

To continue the book metaphor, think of an organism's DNA as a very expensive, rare edition; it is the original version of an organism's genetic story. If it is changed or damaged (if it "mutates"), the organism may no longer be the same.

RNA is like a photocopy, a "transcription," of paragraphs out of this rare book: it is not the original version, but it contains all the genetic information contained by a short section of the DNA. Unlike DNA, RNA has the capacity to travel: it can move outside the cell's nucleus, into a cell's cytoplasm, or it can travel between cells.

Some recent studies have even shown RNA moving between organisms. Once in the cytoplasm, the RNA can be "translated," converting the genetic code into a language made of different letters, the twenty amino acids that make up proteins. These proteins interact with and can make or modify other proteins, lipids, and carbohydrates—an organism's plumbers, carpenters, and electricians—that carry out the work of a cell.

Here's how it works: Just as it does when a cell begins to replicate itself, a sequence of DNA that is being transcribed first unwinds inside the nucleus. But in this case, rather than replicate into an identical double helix of DNA, one strand of the DNA falls to the side, and the other serves as a template for a new strand of RNA. This "transcript" resembles DNA in almost every respect, except that the RNA contains the nucleotide U (for uracil) instead of T. So a DNA strand of AGCT would be transcribed into an RNA strand of UCGA.

Once the transcript of the RNA from the gene is complete (its start and expression activity determined by the "promoter" at the beginning and its length determined by the "terminator" sequences at the end of the original strand of DNA), the single strand of RNA separates from the single strand of DNA, which then twists up again with its original mate. The strand of RNA (known as messenger RNA, or mRNA, for the protein "message" that it encodes) then migrates outside the nucleus, where it binds to a ribosome and the ribosome begins (with the help of transfer RNA, or tRNA) to "translate" the information the mRNA contains from the DNA into the language of amino acids.

But it gets more complex still; it turns out that the Central Dogma—DNA makes RNA, RNA makes proteins—may have been a bit too dogmatic after all. There are RNAs that influence gene expression that are themselves influenced by small or micro

RNAs. Some scientists estimate that, in the human genome, a third of all genes may be regulated by micro RNAs—amazing given that no one even knew about them twenty years ago. And the growing field of epigenetics has shown that gene expression can be determined not just by genetic information alone, but also by stored chemical influences from *outside* a cell, or even outside the organism. (Some research suggests that even mental health and stress affect an organism's genetics and its offspring via epigenetic mechanisms, though there is still much to learn about this.)

Understanding the science of the genome and genetics is vital to the GMO debate. Genetic engineering is fundamentally different from conventional plant breeding. With conventional breeding, you take pollen from one plant and put it on the stigma of another, and hope for a beneficial outcome. It's a comparatively uncontrolled process, and in some cases, depending on the complexity of the trait you're studying, you don't know if you are going to find what you're looking for.

Typically, when you have a breeding program, you're trying to improve the bearing height of a tree, or the taste or texture of the fruit, but there are a million other things going on at the same time. There's never really *proof* that something will work, there's only *history*—these combinations have worked this way in the past. But within this process there remains a degree of mystery. Flower color and disease resistance can be predicted pretty accurately, because they're controlled by just one or a few genes. But yield and adaptation to environment—these are much more complex because they involve multiple controlling genes and are influenced by environmental conditions. This work requires many more plants and growing them in numerous locations.

Exploring these complexities, especially at the level of small

RNAs, is what interests Blake Meyers. Working with colleagues from Stanford and UCLA on a sizable grant from the National Science Foundation, Meyers recently helped sequence the small RNAs in the genes of corn anthers, the male reproductive organs in corn plants. He and his colleagues are also creating an "atlas" of the small RNAs in different plant organs, tissues, and cells; they have worked out a spatiotemporal map of these molecules in the anthers to help them understand how, where, and when they develop and function in maize reproduction.

"The work is slow, tedious, and expensive at this point, so we only design the experiments that we think will tell us something really useful," Meyers said. "You never know where the breakthroughs will come for practical purposes. Who knows when we'll find a key regulator that will fix drought resistance or assist with fertility or hybrid seed production? The odds of one lab finding it are low, but multiply that by the hundreds of labs doing this type of work, and I'm optimistic that the field will come up with some important solutions. The rate of discovery is good and picking up speed."

Working backward, then, the science of gene sequencing is the effort of pulling apart the genetic book to examine how its individual chapters and sentences and words are constructed and arranged. How does their order affect the behavior of the organism as a whole?

This is the mysterious world that Blake Meyers has spent his career trying to penetrate.

"Remember, there are millions, billions, or even trillions of DNA/RNA/protein/gene-expression processes under way that led to every bite of food that you eat, from every plant or animal that has been consumed in the history of the world," Meyers told me. "I would say all but a relative handful of those have never been studied, perhaps never even characterized, and perhaps vary from one bite of food to another."

Meyers and his lab assistants begin sequencing genes by taking a leaf from a soybean plant, for example; freezing it in liquid nitrogen; then grinding it up to break up the leaf's cells. Then, using a series of chemical processes that shatter cell walls, the team extracts the plant's DNA or RNA, and loads them—in chains of fifty to two hundred nucleotides—into a three-inch "flow cell," a kind of glass slide with eight hair-thin capillaries running through it. The slides are then snapped into an Illumina HiSeq gene sequencer.

Depending on whether the team is working on fragments of DNA or short strands of RNA, Meyers's team then needs to code their samples by pumping tiny chains of nucleotides through the flow cells, where they are "amplified" in a series of chain reactions that replicate the sequence millions of times.

Fluorescently charged nucleotides are hit with laser light and photographed at four different wavelengths, showing up on a computer screen as innumerable pinhead-sized "spots" of DNA—the spots that, to my eye, looked like clusters of stars. The microscopic scale of these spots is hard to comprehend. The images are actually measured in microns—that is, each spot is a thousand times smaller than a millimeter—yet each one contains up to 5,000 copies of the same short fragment of genes. And in any series of images, there might be 200 million spots, each one containing a different version of the original piece of DNA or RNA. On each slide, the instruments used by Meyers (and his computers) can detect 1.2 billion sequences of DNA fragments.

The process requires ten terabytes of computer space for each two-week run of the instrument, about what it would take to store the entire printed collection of the Library of Congress.

"Sequencing a genome is like putting together *War and Peace*," Meyers said. "It's long, and content rich, but these machines can only get you words or paragraphs at a time. The machine generating shorter reads of DNA can find you all the times the word 'the' appears, so then it's up to you to find all the places where 'the' appears in the book. But some machines can now give you sentences or entire paragraphs, so you can make much quicker sense of the whole.

"This is like making millions of copies of *War and Peace*. The computer can overlap the fragmented text and look for places where Paragraph 1001 ends and Paragraph 1002 begins, and repeat that process millions of times, analyzing millions of short stretches of words, and incrementally assembling sentences, paragraphs, and whole chapters. Eventually this can help you figure out the sequence of the entire text."

Meyers's team needs so many copies because, statistically, there is always the chance they might find a glitch—a missing paragraph (or ten paragraphs). The longer the strings of text, the easier it is to get recognizable, repeated, and accurate strings.

"As biologists, we're designing experiments to take advantage of these sequences to look into solving practical problems, like drought resistance," Meyers said. "What genes are expressed when plants are distressed? We can compare stressed plants to non-stressed plants. These transcripts are like short-term memory. If the plant were a computer, we'd want to know: Under these conditions, what software has it been running? It's like asking an iPhone, 'What apps have you been using to solve the problems you faced today?' If we looked at your iPhone and saw you'd been using the Lonely Planet app for New Delhi, or looking at the Urban Spoon app, from that information, from that software, we could tell what you were up to. It's the same with plants, looking at their patterns of gene activity."

Tinkering with the Genetic Machine

If understanding a genome is like learning a language, Meyers told me, then using genetics to work on plants is like using a repair manual to fix a faulty wire in a car.

"A cell is like a big machine," Meyers said. "Genes are the blueprints, and proteins are like the different parts of a car. So let's say you have a warning light in your car, and that light means you have a loose wire. You touch the wire and it shocks you. It's not healthy to have that loose wire. In the old days, you would go in there and add a plastic cap. The anti-GMO argument is that this is not the same car; it's been modified by this protective cap. Even though the phenotype itself is better—it's preventing you from getting shocked.

"There are new methods that are more like this analogy: 'Let's go in there with a tool, and remove that wire altogether, or unscrew that bolt. Now you've got the car, but removed the offending wire, and you've taken the tool with you. Is that modified car something that you are unhappy with? It's better than having that protective cap.'"

In the early days of genetic engineering, scientists used .22 caliber rounds—a "gene gun"—to literally blast gene sequences into a plant's cells. Scientists would coat tiny particles of gold with thousands (or millions) of copies of a specific gene sequence they knew would confer a phenotype of interest (for example, making a plant resistant to an herbicide like glyphosate). They would then shoot the gold particles into a group of plant cells. The chances that the gene sequence would actually integrate into the genome in a way that was functional were less than a million to one, but do it enough times (or with enough copies of the gene) and you'll eventually find one cell that lives and has an integrated and functional copy of the gene.

The group of cells would then be treated with glyphosate (aka Roundup), and all of them would die except for a select few. Once a surviving cell was found to have absorbed the foreign gene sequence, it would be allowed to recover and encouraged to proliferate. These cells, now structurally Roundup resistant, could be regenerated by a process of tissue culture into an entire plant.

Meyers doesn't use the gene gun in his lab. Like most of his peers, he uses a bacterium (known as *Agrobacterium tumefaciens*) to do the work instead. With a natural ability to insert their own DNA into plant cells, this soil bacterium can be outfitted with gene sequences scientists want to see integrated into a plant's own genome. Plants are dipped in a solution full of the bacteria, then covered up. The bacterium, a plant pathogen, is effective at infecting the plant and will find its way into the plant stem cells, transferring the foreign DNA into the plant genome just as it evolved to do, and creating a stable transgenic plant.

In his laboratory's "green vaults"—growth chambers that resemble walk-in food coolers—Meyers showed me plants maintained with high humidity and variable light and temperature. Inside were hundreds of examples of two plant species that have become the workhorses for plant geneticists: tobacco and Arabidopsis, the latter a member of the brassica family that includes things like broccoli.

These plants are like lab mice for plants—they flower, they reproduce, they respond to stressors. And the research Meyers is doing is similar to the basic research biomedical researchers do with mice: both are trying to unlock the mysteries of the ways organisms function.

The Arabidopsis genome is a fraction—perhaps 5 percent—the size of the corn genome. It is small enough to grow several plants to maturity in a single coffee cup, which makes it much easier (and cheaper) to work with than corn, which is happiest in a field, well separated from its neighbors. And since "core responses" are similar across all sorts of plants, scientists can play around with an Arabidopsis plant and its genome and extrapolate conclusions that will likely hold true for corn or soybeans.

"Let's say you take two varieties of Arabidopsis," Meyers said. "One grows well in moist climate like Germany. One grows in dry climate like Utah. You can make a cross between them, then use the progeny and traits segregating in those progeny to map onto the chromosomes the loci controlling responses to these climates. You can then identify the locations in the genome that contribute genes important in the line for Utah for drought resistance that may be missing in the line from Germany.

"Say that you've found gene X that confers the ability to survive in drought conditions. As a geneticist, to test that function and demonstrate causality, you may want to break gene X, to see what happens when the plant normally happy in Utah loses the gene you think contributes to fitness under dry conditions. Or you can misregulate gene X, or modify key parts of the protein that gene X produces. With the resulting data, we can make insights into how gene X functions and confers the phenotype that had attributed to it through a standard genetic approach. For all this, your work is greatly facilitated by having a plant into which you can easily introduce the gene. With Arabidopsis, you have a generation time of eight weeks, lots of molecular tools and preexisting data, and great toolkit for molecular biology. It's really easy to work with, particularly relative to most crops."

THE WORLD INSIDE a plant laboratory is so ordered, so controlled, that scientists can be forgiven for their frustration at the screaming debate that has developed over the work they do. "Forty years ago when I was in school, we'd see farms using chemicals to prevent diseases, where the soil was sterile because of the methyl bromide used as a fungicide and biocide for strawberries," Jim Carrington of the Danforth Center in St. Louis told me. "We would visit strawberry fields in California and say, 'Wouldn't it be great if there was a different way to do this, to use a plant's own genetics to fight off diseases more effectively, so we wouldn't have to douse them with fungicides—some of which do have adverse human health effects? And wouldn't it be great if we had plants that help build the soil, rather than forcing us to lose carbon and other organic matter?' We'd imagine we could improve plant genetics, and now some of that has come to fruition. We have low- (or no) till farming, doing all those things that fall under desired practices to promote sustainability. As a scientist who envisioned many of those things— to see all those aims co-opted by groups that are antagonistic to science—is really frustrating."

Scientists are trained to discuss data, not to make political arguments, Blake Meyers said. "We aren't trained in arguments about whether you prefer organic or nonorganic food," Meyers said. "If you tell me this gene is bad for you, we can have that discussion. Show me the data, I might say. But those aren't the arguments that we are often having. It's like a discussion about evolution versus intelligent design. There is no science supporting intelligent design, so to a scientist this debate will be fruitless, as it's irrational. There's no common ground between the rational scientist and the passionate believer. With GMOs, you get the same sort of situation.

"On the public side, arguments against GMOs are often not grounded in science," he said. "They may be based on 'the way things should be,' or 'the way things used to be,' or based on someone's individual opinion. This can be very frustrating for scientists. Those are arguments that don't fit the data. Show me an argument based on data, and we can have a reasonable or at least scientific discussion."

The frustration felt by Carrington and Meyers—that arguments mounted by anti-GMO skeptics are not based on data or are even "antagonistic to science"—is a common refrain that, to the ears of skeptics, sounds ideological in its own right. Scientists have a way of claiming that their field—or the scientific method itself—is somehow beyond reproach. But no matter the natural laws they seek to discover, scientists are people too—and are thus hamstrung by their own preconceptions, desires, and intellectual parameters. John Fagan is a molecular biologist whose professional change of heart on GMOs has made him a leading figure in the scientific debate. In 1994, Fagan became so concerned about the direction of genetic engineering he returned more than $613,000 in grant money to the National Institutes of Health. He quit an academic job to found the Global ID Group, which developed tools for testing genetically modified food, and now directs Earth Open Source, a leading anti-GMO clearinghouse whose publication *GMO Myths and Truths* has become a bible for GMO critics.

"There are bona fide scientists who are doing genetic engineering of crops in one way or another and they really sincerely believe that there is not a problem," Fagan said. "You can get all the way through your PhD without ever having a course in the philosophy of science, or a course that discusses the social or environmental impact of technology. The training is very focused on technical aspects of doing molecular biology in one area or another, and as a result you end up with scientists who are really experts in their own area but oftentimes do not understand the relationship between their work and the world out there. Many of them have the attitude that it would be compromising for them to think about or be involved in a debate about larger issues. They feel that they need to be *scientific* about what they do, and impartial, and true to numbers they get in a lab. There is some merit in that, but on the other side, to have only that perspective on whether a technology is commercialized on a large scale in the world is a very risky thing."

The safety of genetic engineering is not nearly as settled as the majority view claims it is, Fagan maintains.

"The evolution of the debate on GMOs has really evolved over the last twenty years," Fagan said. "Early on we were saying, 'Based on what we know about how genes function, and what we know about the process, we feel this is a very sloppy and imprecise process that could lead to unexpected problems.' Today, there is lots of evidence that says GMOs do *not* function the way we predicted, and there are a lot of unintended side effects that have come up."

The prevailing idea, the Central Dogma—that inserting a single gene into the DNA of another organism will cause a single, predictable change in a single protein, followed by a single change at the cellular level of the organism, a single change at the tissue and organ levels, and a single change at the level of the plant as a whole—fails to recognize the complexity and interconnectedness of the many components of living organisms, according to Fagan.

It is now thought that most genes encode not just one protein but two, three, four, or more, and that the regulatory sequences associated with one gene can influence the expression of neighboring genes. When a new gene is inserted into the DNA of an organism, that gene is likely to influence the expression not just of one gene, but several. Likewise when the newly inserted gene is expressed as a protein, that protein will not have just a single effect, but several. It will influence multiple cellular processes and, subsequently, multiple processes in tissues, organs, and the entire plant. In other words, instead of a single, predictable effect, the insertion of a single gene can result in multiple effects, which can themselves affect many other processes, from the cellular level on up. The more effects, the more unpredictability.

"There are spatial and temporal aspects of this," Richard Manshardt, a plant virologist who helped develop the GM papaya in Hawaii, told me. "The idea that a gene occupies a particular location, and makes a particular protein, and that protein has a single function, is long outdated. Genes are complicated, and they can interact with different parts of the genome in different ways and at different times. It's a much more dynamic system, and this is even before RNA. This is just in the coding sequence of DNA. If there are two functional units, one might interact with a different gene on a different part of the chromosome. For sure, science is always finding out how ignorant we are."

And consider that there are over 20,000 genes in the human genome, but in excess of 200,000 proteins—yet only a small fraction of DNA actually codes for proteins. What is the rest of DNA doing? We don't really know. What was once considered "junk DNA" is only now beginning to be understood—which is further reason for caution when spreading engineered genes across the globe, Fagan said.

"My belief is that nature is parsimonious in what it does," he said. "It doesn't waste anything. There are those hubristic opinions that say, 'If we don't understand it, it doesn't exist, or it's superfluous.' But that's the kind of thinking that allows people to be comfortable with the idea of going in and manipulating new genes in very sloppy ways and being so confident in putting them on millions of acres, and for decades. That kind of logic is really risky."

Arguments about GMO technology are one thing, in other words. The real anxiety arrives when the technology is applied systematically, across wide swaths of the continent and the globe—almost all of it in the service of industrial food. A big part of the problem is that a great deal of university science is funded by industry, or by a federal government in full-throated support of industry, Larry Bohlen told me. Bohlen is a veteran environmental activist who made international headlines fifteen years ago when he discovered that a GM corn (known as StarLink and approved only for animal feed) had made it into the human food supply.

Consider the funding that flows from the USDA into research on GMOs. "When I looked at it in 2002, it was $193 million for GMO research, of which $3 million was to look at potential environmental problems—or about 1.5 percent—and zero for health effects," Bohlen said. "That would be $2 billion over the last ten years. If I want to survive in academia, of course I'm going to go after the piece of the pie that is 98.5 percent of the budget. There's no real blame in that—you can just look at it objectively and see that most of the money is going to the promotion of GMOs, so that's where the scientists are going. And somebody is setting that budget."

Blake Meyers is matter-of-fact when it comes to the funding of his—and all—scientific research. His work is supported by the NSF and the USDA, but he has also gotten money from the big industrial players: DuPont, Dow, Syngenta, BASF. The building that houses his lab was constructed in the mid-1990s, with funds from the University of Delaware, the state of Delaware, and DuPont, whose world headquarters are just up the road in Wilmington. The institute was designed as a hub for research and teaching in the life sciences, but also to support the development of start-up biotech companies.

"Thirty years ago the university got 40 percent of its budget from the state," Meyers said. "Now it gets 12 percent. The funding rate for federal grants has declined significantly as well. As these sources of funding have declined, there's been a push to diversify the sources of funding for research. So as scientists, you have to find a way to support your lab and support your graduate students. I'm a basic research scientist. Industry comes to me, and they don't give me money for science because they think I'm a great guy. They say, 'We don't know how to measure small RNAs, and we need some help.' My academic group has expertise that can help them to accomplish their goals, and academics can also learn from projects with our industry colleagues."

This dynamic—the drying up of publicly funded research, and its replacement by research paid for by industry—naturally results in an agricultural system dictated by industry, critics say. But the larger issue has to do with the way this thinking—manipulating genetics to serve industrial purposes—has changed the way scientists see the world. Cells are not really like machines, as Meyers suggests, and tinkering with genes is not really like tinkering with a car engine, Craig Holdrege, a scientist and philosopher who runs the Nature Institute in New York, told me. Like organisms out in the world, genes operate in dynamic systems, and both context and timing are far more complex than most scientists allow.

"In a way we're treating organisms as if they were made up of independent parts, and you can put things in or exchange them, and come up with a result that you (as a human being) like," Holdrege said. "You think you have that degree of control, and you can manipulate that organism to do what you want it to do. But if you read in the literature of genetics and epigenetics, it's completely clear that context matters, that timing matters, and you cannot say that there is a very particular 'this' that always causes a particular 'that.' That's what we've all been indoctrinated to think. Even though the literature is screaming that at us, the habit of thought about causal mechanisms is very deeply entrenched and not easy to overcome, or to move beyond, or to see its limitations. We need to take more seriously the fluidity, the plasticity, and the interconnectedness of all structures and processes."

How you feel about GMOs, whether you are a consumer or a biologist, may have less to do with your grasp of complex science or tangled agricultural history and more to do with how you view your place in the world. Yale University's Dan Kahan recently asked more than 1,500 Americans to rate the threat of climate change on a scale of 0 to 10, then correlated their responses with their scientific literacy. He found that higher literacy was associated with stronger views at *both* ends of the spectrum. Science literacy, in other words, promoted polarization, not consensus. People use scientific knowledge to reinforce beliefs that have already been shaped by their worldview.

Similar passions have polarized the country on GMOs, in part because so many of the issues are the same as they are with climate change: big corporations, big government, big fears.

Americans fall into two basic camps, Kahan says. Those who think of themselves as "egalitarian" or "communitarian" are generally suspicious of industry, which they would like overseen by regulators. In contrast, people who see themselves as "hierarchical" or "individualistic" respect leaders of industry and dislike government interference, which they presume would lead to taxes or regulations. Take a barber in a rural town in South Carolina, Kahan writes. If his customers were skeptical about climate change, would it be smart for the barber to urge them to petition Congress to limit industrial emissions? If he did, Kahan writes, he would find himself "out of a job."

When we argue about GMOs, or climate change, in other words, we are also arguing about who we are and what our crowd believes. "It's fascinating, almost mesmerizing, how personally involved we get in these things," Richard Manshardt, the Hawaiian papaya researcher, told me. "We don't want to be manipulated. We want to control our destiny, and sometimes that means doing absolutely stupid and irrational things. People think, 'GMOs are bad because the group I'm with doesn't like them,' and it ends up being all about my standing within my group. Your own logic, your own sense of what's right, gets challenged, and most of us are not comfortable with changing that. We are not open to new kinds of challenges."

Of course, this works both ways. It's not just members of the "unscientific public" who fear being seen as outliers in their own social group. It is scientists too. Indeed, in its intensity—as well as the size of the stakes—the scientific debate over GMOs has become almost theological, Brian Snyder, the head of the Pennsylvania Association for Sustainable Agriculture, told me. "We're not just talking about data, we're talking about which worldview to fit data *into*," Snyder said. "It's not just this side and that side. People use the same science and reach different conclusions. At the very least, scientists are making subjective decisions about what to study, and often their results will follow from this subjectivity.

"It's fascinating to me that so much of the technology is being developed in order to address problems caused by the previous technology," Snyder said. "So when you boil that down you realize there's sort of an undying faith that answers are always going to come out of the laboratory, that we're just one discovery away from solving it all."

Heisenberg's uncertainty principle suggests that a scientist's worldview has a great deal to say about what he will get from his data, Snyder said. "We really are right down to pointing out world issues, paradigm issues. We have no effective way of carrying on a conversation about conflicting worldviews. We used to chop the heads off people who said the earth revolves around the sun."

The Séralini Affair

If ever there was a moment when scientists were ready to chop off some heads, it came after a journal article claimed to prove that GMOs—and the common herbicide glyphosate—caused cancer and premature death in rats. The study, by Gilles-Éric Séralini of the University of Caen in Normandy, France, monitored two hundred rats for two years. The rats were divided into ten groups, each with ten males and ten females. Some groups were fed different amounts of a strain of Monsanto corn (called NK603) that had been engineered to resist glyphosate. Some corn had been sprayed with glyphosate, some had not. Other groups were fed glyphosate in their drinking water. There was also a control group, which was fed non-engineered corn and plain water.

Séralini found that female rats that ate both the engineered corn and the glyphosate tended to develop mammary tumors and compromised sex hormones—and tended to die earlier—than the rats in the control group. Male rats showed four times as many large, palpable tumors and "very significant" kidney deficiencies.

The study, which passed the traditional peer-review process, was published in 2012 in the journal *Food and Chemical Toxicology*, one of the leading publications in the field.

It went off, in the words of France's environment minister, like "a bomb."

Jean-Marc Ayrault, France's prime minister, said that if its results were confirmed, then his government would press for a continental ban on NK603 corn. Russia suspended imports of the crop. Kenya banned *all* GM crops. The article appeared two months before a referendum in California that would require the labeling of all GM foods.

In other words, a single article threatened to tip the global conversation about GMOs, and not in the food industry's favor.

Almost instantly, the journal was deluged with letters savaging its conclusions. Critics complained that the experiment used too few animals; that the rats used in the experiment were prone to cancer anyway; that the experimental protocol used could not distinguish tumors caused by GM food from those that would have occurred anyway. It was "clear from even a superficial reading that this paper was not fit for publication," a professor at the University of Cambridge wrote. "The study appeared to sweep aside all known benchmarks of scientific good practice and, more importantly, to ignore the minimal standards of scientific and ethical conduct in particular concerning the humane treatment of experimental animals," a group of prominent scientists concluded.

Citing such "major flaws," industry officials and pro-GMO scientists alike called for the article to be retracted. In an extraordinary and highly unusual move, the journal's editor complied.

"Unequivocally," the editor wrote, there was "no evidence of fraud or intentional misrepresentation of the data." Nonetheless, there was "legitimate cause for concern regarding both the number of animals in each study group and the particular strain selected."

A more in-depth look at the raw data "revealed that no definitive conclusions can be reached with this small sample size regarding the role of either NK603 or glyphosate in regards to overall mortality or tumor incidence," the editor wrote. "Ultimately, the results presented (while not incorrect) are inconclusive, and therefore do not reach the threshold of publication for *Food and Chemical Toxicology.*"

The decision to retract caused a second firestorm that was at least as powerful as the first. More than a hundred scientists signed a petition calling the decision "arbitrary" and "groundless." Retracting a published, thoroughly peer-reviewed paper "is without precedent in the history of scientific publishing, and raises grave concerns over the integrity and impartiality of science."

Among other things, critics noted, was the fact that Séralini had used the same strain of rats Monsanto had used eight years earlier in its own study, which persuaded European regulators to approve the use of GM corn in the first place. That study—done over ninety days, compared with Séralini's two years—had been published in the same journal that was now retracting the new study.

"The retraction is erasing from the public record results that are potentially of very great importance for public health," the petition said. "It is censorship of scientific research, knowledge, and understanding, an abuse of science striking at the very heart of science and democracy, and science for the public good."

The decision was based "not on the grounds of fraud, malpractice or data misrepresentation, but simply (as far as I can see) because Monsanto and its legion of followers did not like the results of the research and have given you a hard time," wrote the British environmental scientist Brian John.

"The campaign of synthetic outrage orchestrated by the GM industry against the paper and against Séralini personally was something that the scientific community should be thoroughly ashamed of, since it was characterised not just by a lack of respect for the research team and its findings, but by personal vilification the like of which I have not seen for a long time," John wrote. "'Inconclusiveness' is not a ground for retraction—every scientific paper published is inconclusive in the sense that it might show probability and might point the way for future research. That is exactly what the Séralini paper does, in a perfectly responsible way. If you press ahead with this, you will also confirm what many people have been increasingly concerned about—corporate control not only of the biotechnology industry but also of the means of publication. That is both scientifically reprehensible and sinister."

Not long after the retraction, the journal further infuriated critics by installing Richard Goodman, a former Monsanto scientist, as its new associate editor for biotechnology. Goodman's "fast-tracked appointment directly onto the upper editorial board raises urgent questions," an article in *Independent Science News* reported. "Does Monsanto now effectively decide which papers on biotechnology are published in *FCT*? And is this part of an attempt by Monsanto and the life science industry to seize control of science?"

Brian John offered a sharp answer for this as well. "Only a fool would assume that there is no connection between his arrival and this decision to retract the Séralini paper," John wrote. "And from where I stand this is yet more evidence of the increasing corporate control exerted by the GM industry in the area of biotechnology publications. You clearly do not now have true independence in editorial matters, and the manner in which you have buckled under pressure from this orchestrated anti-Séralini campaign is both despicable and deeply depressing."

In an open letter in *Independent Science News*, a group of scientists wrote that the Séralini affair risked undermining the credibility of science itself:

When those with a vested interest attempt to sow unreasonable doubt around inconvenient results, or when governments exploit political opportunities by picking and choosing from scientific evidence, they jeopardize public confidence in scientific methods and institutions, and also put their own citizenry at risk. Safety testing, science-based regulation, and the scientific process itself, depend crucially on widespread trust in a body of scientists devoted to the public interest and professional integrity. If

instead, the starting point of a scientific product assessment is an approval process rigged in favour of the applicant, backed up by systematic suppression of independent scientists working in the public interest, then there can never be an honest, rational or scientific debate.

The issues raised by the Séralini study have not gone away. In a move that may one day generate as much noise as the Séralini affair itself, the Russian National Association for Genetic Safety, a nongovernmental, nonprofit organization based in Moscow, said in 2014 it was raising $25 million to redo Séralini's experiments. In a theatrical press release (announcing the *launch* of the study rather than its results), the designers of the so-called Factor study said the GMO controversy was so hot they would not even disclose where the research would take place.

The Factor study will use 6,000 rats, rather than the 200 used by Séralini; last four years instead of two; and examine the health consequences of both GM food and glyphosate as they manifest through multiple generations. It will adhere to or exceed guidelines set by the Organization for Economic Cooperation and Development (OECD), an international research body that works with governments on economic and environmental policy, said Bruce Blumberg, a professor of developmental and cell biology at UC-Irvine, who will serve on the study's scientific oversight board.

"The study will employ large groups of animals tested throughout their full lifespan, so unless you are Monsanto, the results will not be easily disputable," Blumberg told me. "The work will be very thorough and very transparent. For sure, companies don't do that. They never let their data out."

Such independent science is critical to the GMO debate, Blumberg said, since most of what we know about the safety of

glyphosate—the herbicide most commonly sprayed on GM crops—
is based on studies Monsanto did back in the 1970s and 1980s.

"We remain reliant for what we know—about a chemical sprayed
on hundreds of millions of acres of food crops—on studies performed
by the company that produces and sells it," Blumberg said. "The
bottom line is that we know almost nothing about the long-term
health effects of this compound and we use it in colossal amounts.
If you talk to Monsanto, they will say, 'If you use our seeds, you
will use less Roundup,' but the twentyfold increase in the use of
glyphosate-based herbicides since 1996 disputes this assertion."

Blumberg's own work has focused not on genetic engineering
but on the effect on human health caused by hormone-disrupting
chemicals. He looks at the ways synthetic chemicals contribute to
things like cancer and obesity. "Obesogens," such as those found in
certain synthetic plastics and fungicides, for example, can cause a
body to make more fat cells, to put more fat *into* cells, or to change
a body's overall metabolism. Any of these disruptions can cause a
rat—or a human—to put on weight, regardless of their diet. Significantly, these chemicals can cause such changes not just in
exposed individuals, but in their offspring down at least three generations.

Looking over one of Blumberg's research grant proposals ten
years ago, a reviewer at the National Institutes of Health wrote,
"How dare you waste our time with such a ridiculous idea—everyone knows obesity is caused by eating too much and exercising too
little." Now, thanks in part to Blumberg's work, we have come to
know much more about the way synthetic chemicals affect our
body's subtle hormonal balance—and how easy it is for them to
throw this balance out of whack. National trends—from obesity to
early-onset puberty in girls or low sperm counts in boys—have been

traced to the chemicals used in an enormous variety of everyday consumer products.

Blumberg, in other words, may not have a horse in the GMO race, but he's seen similar contests before.

"My job is to make sure the study is designed as well as it can be," he said. "The goal is to see if we can test the safety of one of the world's most widely used chemicals. As a human being who inhabits this planet, I'm very interested in seeing this done well. I want to see the results."

Part Two

SEEDS

The Fruit That Saved an Island

The vast majority of GMOs are hidden in highly processed ingredients in the supermarket's meat and junk food sections, but there is one modest little GM fruit you can find in the produce aisle. The Rainbow papaya, grown on the Big Island in Hawaii, is a singular retort to the blanket condemnation of genetic engineering. The fruit is nutritious. It nurtures local farmers. And it was created not by a company, but by a professor.

Dennis Gonsalves unlocks the metal gate blocking the entrance to the 80-acre plantation, then returns to the driver's seat of his beat-up white Ford pickup. We pass field after field of scrawny trees, their feet, here on the Big Island, growing in volcanic rock, their crowns bursting with fruit.

Moments later, we pull up to a large, open-air enclosure lined with plastic crates, filled with freshly picked, genetically modified papaya. Gonsalves has had a long love affair with this fruit. He should—he designed it himself.

At seventy, Gonsalves is physically robust, ebullient, and remarkably charismatic. In the world of genetic engineering, he has long been considered something of a superstar: a university scientist who saved an industry, and who did it without buckets of government or corporate money. Gonsalves, many people say, is the very model of the way science should be done: with the public good, not corporate profits, at heart.

As he parks his pickup, Gonsalves calls out to the plantation's owner. "Hey, Alberto!" Gonsalves shouts as we leave the truck.

Alberto Belmes climbs down from a forklift and comes over to shake hands. Belmes came to Hawaii from the Philippines in 1981 and started planting papaya a year later. Within ten years, his fortunes, and the fortunes of the trees he was planting, were in danger of complete collapse. A pathogen known as ringspot virus was burning through the trees on the Big Island, and neither farmer nor spray gun had any way to stop it. Ringspot had destroyed the papaya crop on Oahu in the 1950s, prompting the entire industry to move to Puna, on the Big Island, where the virus, up until then, had not yet manifested.

By the early 1990s, ringspot was moving across the Big Island like a wildfire. State agriculture workers did their best to destroy infected trees, but nothing could stop the burn.

Like Belmes, 80 percent of the island's papaya farmers were first-generation immigrants from the Philippines, and few of them spoke English. They were poor, they had few political connections, and they were dependent on their plantation work to feed their families. The ringspot virus was not just killing trees, it was threatening to ruin a fragile human population as well.

At first, Belmes tried to outsmart the pathogen by planting his trees close to the ocean, where salt air and wind kept many of the aphids spreading the virus at bay. Early on, as other farmers' crops crashed, Belmes's trees were still producing fruit, and he was able to charge a premium: he sold papaya for a dollar a pound, three times what the fruit would command twenty years later.

Flush with cash, and keeping his fingers crossed, Belmes took a risk. He planted another seventy acres and began moving his crop back inland. Maybe the virus had moved on. Maybe he'd get lucky again.

His plan failed. His newly planted trees were devastated. That year Belmes didn't harvest a single fruit. He lost everything.

"People forget how bad it was," Dennis Gonsalves told me, casting his eyes across the volcanic landscape. "It was like a war zone here. All the trees were dead."

THE PAPAYA is a remarkable plant: a giant herb, really, not a tree. From the moment a seed is planted, it takes just six months for the plant to grow several feet tall and begin flowering, and just six more months to begin producing fruit. After three years, a papaya tree can be eighteen feet tall, producing scores of fruit at a time.

Dennis Gonsalves is a remarkable man. He was raised on a Big Island sugar plantation, where his father worked mowing grass. Times were tough: he remembers his family and the rest of the workers eating a lot of Spam. "The bosses were white people and the workers were locals," he said. "I grew up with that mentality."

Gonsalves left the island to get his PhD in plant pathology at UC-Davis, then taught at the University of Florida for six years. Bored of spending all his time studying citrus, he moved to Cornell, in New York, where he would spend the next twenty-five years working alongside some of the smartest plant pathologists in the business. As a plant virologist who knew how bad ringspot had been on Oahu, Gonsalves started experimenting on papaya in 1978. He traveled all over the world trying to figure out how to protect the crop from viral pests. This was before genetic engineering was developed, so Gonsalves and his research team tried to protect plants by "immunizing" them to protect against serious infection.

Initially, Gonsalves tried a kind of vaccination called cross-protection: he and his team mutated a mild strain of the ringspot virus and used it to inoculate millions of trees. Plants do not have active immune systems, and thus can't produce antibodies to protect themselves against diseases. Plant biologists had long known, though, that when they are exposed to a weak virus, plants can develop some resistance.

Cross-protection had drawbacks, however. There was always the chance that the mild virus injected into the trees might mutate into a far more damaging form. Or the virus protecting the papaya could jump species and cause serious infections in another important agricultural crop.

By 1983, it was clear that cross-protection was working, "but not that well," Gonsalves told me. So, mid-career, the veteran scientist started teaching himself some new tricks.

The mid-1980s was an exciting time in the field of molecular biology, and Gonsalves didn't have to look far for an example of a promising experiment. Washington University's Roger Beachy had recently managed to take a strand of DNA from the tobacco mosaic virus—a common pest that damages a plant's leaves and can stunt its growth—and insert it into the tobacco plant itself. The strand of virus DNA they used triggered the creation of "coat proteins," which—in the virus—served as a kind of shield against outside infection. When Beachy inserted this viral DNA into the tobacco's DNA, it did the same thing: it protected the tobacco plant from infection—in this case from the very virus that had contributed its DNA to the plant. Though the process remained somewhat mysterious, one thing was clear: the transformed plants were resistant to infection from the tobacco mosaic virus.

Following Beachy's lead, Gonsalves and a colleague, Richard Manshardt, wanted to see if they could pull off the same trick using ringspot virus and Hawaiian papaya. They knew how ringspot worked: it hijacked the papaya cell's protein-making machinery. They started experimenting with gene sequences that might cause RNA interference; they wanted to get their papaya to produce a small stretch of genetic code that could, in turn, spur a biochemical process that would seek out and "silence" infecting strands of viral RNA. Once attacked, they hoped, the viral RNA would be rendered "mute."

Meanwhile, a Gonsalves colleague at Cornell, John Sanford, was developing the "gene gun," with which he learned to shoot DNA-coated tungsten balls into plant cells. It was an imprecise method, to say the least; only a tiny fraction of the target cells would absorb and incorporate the new DNA. But some DNA did make it into the target cells, and the possibilities for genetic manipulations suddenly opened up.

The mystery of cell biology—combined with the prospect of real crops facing real disaster—made Dennis Gonsalves's genetic work both intellectually exciting and economically urgent. Given the pending collapse of the papaya industry, there just wasn't time to plod through years of traditional breeding experiments.

For the papaya plantations on the Big Island, these experiments could not have come at a more critical time. In 1992, a dean at the University of Hawaii at Manoa called Gonsalves to break the news that the ringspot virus—once absent from the Big Island—had popped up in Hilo, just twenty miles north of Puna, where virtually all of Hawaii's commercial papaya crop was being grown. Gonsalves had discovered his professional calling. "It is rather rare that a potential solution is coincidental with a potential disaster," Gonsalves wrote at the time, with considerable scientific understatement.

"I was new at Cornell, but 95 percent of Hawaii's papaya industry was here on the Big Island," he told me later. "I started to realize that good science could change things at home."

At first, Gonsalves's research did not attract much attention—or money. His initial grant application was rejected by both the USDA and the NSF; small grants came thanks to the reach of Hawaii's U.S. Senator Daniel Inouye, who knew the state's economy would take a significant hit if the papaya industry collapsed again. In what has become something of a legend among the world's plant geneticists, there was no corporate money involved.

"The first grant we got was for $20,000," Gonsalves said. "This was a poor man's biotech project. We were just scientists doing research."

Not all of the hurdles were scientific. Gonsalves also had to seek approval from the entire range of the federal bureaucracy: the USDA, the EPA, and the FDA. The research team conducted numerous toxicity tests and protein studies, some of which would never have been required of plants grown in a traditional breeding program. In this, as in its status as one of the earliest GMO experiments, Gonsalves's papaya became a harbinger of future battles.

"When the Rainbow papaya was first engineered, its developers had to test several kilos of dried papaya leaves to see if the alkaloid content was any different from traditionally bred fruit. It wasn't," Richard Manshardt told me. "That sort of thing would never be looked at in a conventional breeding project.

"The big difference between traditional breeding and GMO, people think that what we eat now is 'natural' and has been around for centuries. Their experience is okay, and therefore the government isn't under any pressure to impose further testing. But with

GMOs, it's a new process. The government requires researchers to do the testing, and the public sees the whole system as suspicious and untrustworthy. Why? Because it's new, there's potential for a Wall Street distortion of reality, where money is more important than the product. So it's a trust issue, I think. From a research standpoint, the regulatory process is important because we don't know everything, and we *know* we don't know everything."

Yet by the fall of 1997, with a speed that surprised everyone in the field, Dennis Gonsalves had the approval he needed. GM papayas were officially deemed safe for human consumption, and for the environment, by the American government.

Roger Beachy's technology, which Gonsalves was building on, had already been licensed to Monsanto, and Hawaiian papaya growers were deeply skeptical of the biotech giant, figuring the company would charge millions of dollars for them to use company property. But given Roger Beachy's influence—and his strong desire to see his technology work in the field—Monsanto issued the licenses for almost nothing.

On May 1, 1998, after the patent licenses came through, the Rainbow papaya seeds were ready for distribution, and Gonsalves handed them out to island farmers for free.

Alberto Belmes was one of the first five farmers to try them. He planted seven acres. "We didn't know if it would survive, or if people would like it," Belmes told me.

Gonsalves smiled. "We scientists—we were confident," he said.

One year later, Dennis Gonsalves's seeds had turned into Alberto Belmes's trees, and they were bearing fruit. Enormous quantities of fruit. Conventionally grown trees, stunted with yellowed, infected leaves, average just 5,000 pounds per acre per year. The GM Rainbow papaya trees provided 125,000 pounds. From a low of 26 million pounds in 1998, the papaya crop grew to 40 million pounds just three years later. Rainbow papaya seeds are currently controlled by a nonprofit industry group called the Papaya Administrative Committee; seeds are distributed to local farmers at cost.

In 2002, Gonsalves and his research team were awarded the Humboldt Prize for the most significant contribution to U.S. agriculture in the previous five years. Gonsalves is "a tireless innovator," said Pamela Ronald, a plant geneticist at UC-Davis who has done groundbreaking research into genetically engineered rice. "Not only did he return to his home to help the farmers in his area, he moved beyond basic science to getting his invention out in the field. His work is widely viewed as brilliant."

Gonsalves's work "is a model for what should have happened [everywhere]," said Roger Beachy, whose work on the tobacco mosaic virus helped inspire Gonsalves's own work. "He just plain stuck to it because the farming industry needed it."

At seventy, Gonsalves remains undaunted. He is currently trying to open the gates to GM papaya in China, one of the world's biggest markets. He has submitted the scientific paperwork and notified the embassies. Chinese scientists have visited Hawaii's plantations; they've done health data tests and rat-feeding experiments. Gonsalves has received permission to send seeds.

"The Chinese have 40 million people in Beijing and Shanghai alone," he said, "and they *love* Hawaiian papaya. That's what hard work does."

Yet ask Gonsalves what makes him most proud, and he will point to working farmers like Alberto Belmes, the papaya farmer who migrated from the Philippines to Hawaii. Were it not for Gonsalves's discoveries in the laboratory and the field, Belmes and many others like him would not have survived these last thirty-five years. Today, Belmes has twelve employees and close to 100 acres of highly productive papaya trees. He has one son who graduated from New

York University (and now works in a bank) and another son in college in Hawaii.

These days, Belmes spends a lot of time touring foreign scientists around his farm, showing them the benefits of genetic engineering. Recently, a group from Japan arrived and—convinced that Belmes must be suffering because of all the anti-GMO rhetoric floating around the world—asked if he had had any second thoughts about his trees.

What about organic? they asked. Hadn't the global interest in organic farming made GMOs a risky investment?

"I said, 'Ever since the Rainbow came out, it has been very good,'" Belmes told me, chuckling. "We couldn't even supply the market. They thought I wouldn't be happy. They thought GMOs were dangerous!"

Belmes's plantation is now one of the biggest on the Big Island. The day Gonsalves and I visited, a half-dozen pickers—equipped with long bamboo poles and canvas bags—were knocking ripe fruit out of trees.

"Isn't this incredible?" Gonsalves told me, waving his hand across a landscape of thickly planted, highly productive papaya trees.

"I used to come here and cry."

FOR DENNIS GONSALVES, all the noise over GMOs—the politics, and the corporate money, and the anti-technology activism—has proven a long and bitter irony. After retiring from Cornell, Gonsalves was lured back to Hawaii to work in the USDA's Pacific Basin Agricultural Research Center, where he spent ten years continuing to do science in support of local growers.

"I decided to come back and do the highest levels of science, but also to work for local issues, with respect for all the local culture," he told me. "This really transformed the center into a place that could relate to people. We saw that we could really help the farmers. If you do good science and help people, things will fall into place."

But his return home, where local activists were beginning to launch a campaign to ban genetically engineered crops, also reminded Gonsalves of his own decades of squabbling over GMOs. Back in 1986, when he was still doing research on cross-protection, a crisis emerged in the papaya crop in Thailand. Cross-protection had been moderately successful, but as the years passed, the country's papaya crop faced collapse. Desperate, an undersecretary of agriculture approached Gonsalves, pleading with him to experiment with a GM plant.

Gonsalves asked for a scientist and $15,000 for supplies. By 1997, his team took tissue samples and plants to Thailand. They navigated all the bureaucratic quarantine procedures. They conducted "beautiful" field trials and rabbit studies, and were ready to plant. Their scientific work was solid. What they didn't count on was the intensity of global food politics. Corporate GMOs had earned such a toxic reputation that even Gonsalves's nonprofit plant research stirred up a storm.

Activists from Greenpeace, dressed in gas masks and white hazard suits and carrying signs that read "Stop GMOs!" destroyed the Thai field trials. Greenpeace was making a larger point: papaya wasn't the only GM plant scientists were experimenting with in Thailand. Monsanto was also in the country, trying to get permission to plant GM corn and cotton. As evidence of GMO contamination of traditional crops began to emerge, the Thai government began taking a harder line on experimental crops. The government eventually placed a moratorium on biotech crops, and Gonsalves's plants "never saw the light of day."

"Who suffered from that? The farmers," Gonsalves told me. "Greenpeace said, 'This isn't about papaya. It's about opening the door to big companies. What happens if that gate gets opened?'"

Ironically for Gonsalves, this argument soon began bubbling up in Hawaii. Despite Gonsalves's status as an international agricultural star, and despite his GM papaya having become one of the most famous fruits on earth, the GMO debate is far from settled even on Gonsalves's home island. Anti-GMO activists point out that the Rainbow papaya has been shut out of a number of global markets because of local resistance to GMOs. In Jamaica, an experimental crop was shelved because consumers in Britain—the primary market for Jamaican fruit—would not touch it. Japan initially refused to allow GM papaya into its lucrative market, a decision that forced some Big Island growers into bankruptcy. In 2011, after Hawaii spent thirteen years (and $13 million of taxpayer money) lobbying to get the GM papaya into Japan, that country's Ministry of Agriculture finally agreed to let GM papaya in—but only if they are labeled "GMO."

There have also been problems closer to home. By 2004, contamination—by GM papaya trees of non-GM trees—was found to be ubiquitous on the island, forcing even non-GMO farmers to test their trees and fruit before they were allowed to ship their fruit to Japan. Organic farmers lost markets, seed lines, certifications, and chopped down their trees in order to keep their organic integrity, writes Melanie Bondera, an organic farmer in Kona and a cofounder of several anti-GMO groups in Hawaii.

In 2002, the year Gonsalves won the Humboldt Award, some anti-GMO activists approached him and—somewhat incredibly—asked him to reevaluate his life's work.

"They said, 'Come out against GMO papaya and you'll be

considered a savior,'" Gonsalves told me. "But when they said it was unsafe, I said, 'Show me the data.' You say it's bad for the environment, I say, 'Show me the data.'"

Gonsalves has little patience for the GMO debate, especially when he hears people tell him that his beloved papaya should be grown organically. With 100 inches of rain a year, papaya plantations need no irrigation. But so much rain means an endless challenge from fungi—and the need for regular spraying with fungicide.

"Organic? Bah!" he said. "People live in a make-believe world. Organic is what, 2 percent of the food supply? I don't eat organic. Can you grow organic with this rain? We have a fungus problem. All our organic produce comes from California, and it's going to be that way for a long, long time."

When it comes to the anti-GMO movement, Gonsalves told me, talk is cheap. "Farmers are not stupid. They will take the best way they can to make money," he said. "You want us to do things organically and sustainably? Show us how to do it. Don't talk about it. Do it. I'll clap my hands. Wonderful! But *do* it. Don't just talk about it. *Do* it."

Gonsalves now serves on a science advisory board for the Gates Foundation, which is coordinating a great deal of research funding for projects in the developing world. He also continues to consult with global food conglomerates. It's true, the big seed companies have terrible PR, he said; Monsanto continues to pay for its past sins of arrogance, like forcing GM corn on people in Europe.

"Companies always come to me and say, 'What can we do about our bad PR?' I say, 'Do something that will help the people.' Everything about the big companies, this technology—it's not designed to help small farmers. It's designed to help big companies. If Monsanto had come out with a resistant tomato first, instead of corn, things would have been different. But the company knew there were only millions there, not billions. There was no way they were going to do the tomato."

To be fair, he says, Monsanto has put $80 million into a cassava project in Africa, but does not get enough credit for it. This is the same company, he reminded me, that owned a number of patents on the papaya, but "gave them all to us."

It's important for people to realize how important GMO technology can be for small-scale farmers whose livelihoods face collapse without them, Gonsalves said. Monsanto will never get into niche markets like papaya; there's just not enough money in it. It's up to publicly funded, university researchers to work on crops that help small farmers and poor consumers. Many researchers have become too comfortable doing work—often funded by the big companies—on crops that flow into the industrial food system. This sentiment has been echoed by the National Academy of Sciences, which cited the lack of biotech work on specialty crops as one of farming's most pressing problems.

There are signs that Gonsalves's message is getting through. Last year, federal regulators approved a GM plum, developed by USDA scientists, that resists a deadly European pox. Blight-blocking peanuts are under way. In Florida and Texas, scientists are working to develop what is almost certain to become a flash point for consumers: oranges, which are under siege from a bacterial infection known as huanglongbing, or citrus greening. Building on the work Gonsalves did with papaya, and Roger Beachy did with tobacco, scientists are now scrambling to develop GM oranges to block the disease. Public scrutiny, not to say outrage, is sure to follow.

And if China starts buying his papaya, the greatest beneficiaries, Gonsalves knows, will be his beloved plantation workers. "People come to Hawaii to see paradise. They have no idea," he told me. "You know how much people here get paid? Ten dollars an hour. You know how expensive food is?

"If you don't like the companies, say, 'Break them up like they did with the oil companies!' But don't say, 'It's not safe.' Once the Supreme Court said you could patent everything under the sun, that became the law. If you don't follow the law, how are you going to operate a democracy? You want things to be different? Take them to court!

"I'm for the underdog, the poor people," Gonsalves continued. "All I know is that farmers here were suffering. The human side of biotech is missing. This is not an industrial crop. It is family farming. All we were ever doing was trying to help farmers. That's all we wanted to do."

Trouble in Paradise

The island of Kauai is so beautiful it can make you twitch. The great green slabs thrusting up from the central mountains look like they could be hiding another Machu Picchu; the island's lush, rolling piedmont drops into beaches so famous for their waves that locals have been known to remove uninvited surfers with their fists.

Kauai is also a place where you can see a guy dressed up as the Grim Reaper—black cape, flaming red death mask—standing by a major intersection with a sign that says "Monsanto Sucks!" It is an island where anger at giant chemical companies is so intense that a man who is both a professional surfer and a professional mixed martial arts fighter recently ran for mayor on an anti-GMO platform and got 40 percent of the vote.

Tiny Kauai, perched at the far western edge of the United States, has become ground zero for the global debate over genetically modified food and the spraying of their attendant chemicals on cropland. It is a place where, for years, multinational agrochemical companies have developed the GM seeds that circulate around the globe, but kept their experiments—especially their use of pesticides—secret from the people who live just down the road. And it is a place where a ragtag group of activists have fought these companies to a draw. Like other communities around the world that have fought the agrochemical conglomerates, the people of Kauai feel they are bearing a chemical onslaught their bodies and their beloved island ought not to have to bear. They argue that their land is being used for the good of company profits, that GMOs are really just a vehicle for chemical companies to sell the world more pesticides, and that their fight is a microcosm of the global GMO battle itself. Indeed, when it comes to the global food system, with all its perils and promises, the rest of the world is watching Kauai. Because just as GMOs and their attendant pesticides can spread around the world, so can resistance.

When I arrived in Kauai, the guy at the rental car agency asked me why I had come to visit. "I'm writing a book about GMOs," I told him.

"Huh," he said. "Good idea. Lots of pesticides being sprayed over on the island's west side. A guy I work with just lost his dad over there. It's strange how many people are telling stories like that. I'm glad I work inside."

When I pulled into my hotel, the woman at the check-in desk also asked me why I had come. I told her.

"My husband works for a fertilizer company, and he says all this stuff about GMO companies is nonsense," she said. "Closing these companies down would be taking food right out of people's mouths. I just try to keep quiet.

"Be careful who you talk to—you might end up starting a fight."

A few hours later, I found myself sitting in the passenger seat of a beat-up Toyota pickup truck being driven by Jeri DiPietro. We had bumped along an endless series of potholes down a long dusty road, finally pulling up next to an abandoned sugar mill, its exterior walls overgrown with trees and weeds. A pair of rusted-out truck chassis sat rotting in front. Behind them, a squat conical building had been emblazoned with a line of graffiti: "It all started here."

Behind us, across the dirt road from the abandoned mill, a series of squat plastic silos filled with a yellowish liquid sat baking in the sun.

DiPietro had come to this place—an experimental farm operated by the agribusiness giant DuPont Pioneer, to see if she could figure out what the company was spraying on its fields. She carried with her a series of maps showing the locations of company fields, amended with thick lines of Magic Marker that showed acreage and field boundaries (Pioneer 4,500; BASF 900; Syngenta 3,000; Dow 3,500 + 500), as well as their proximity to local rivers and towns and the pesticides being used there.

"There is a field in Kamakani on the west side—all we have is Google Earth to see where the fields are," DiPietro told me. Chemical companies have fields "within 450 feet of a preschool, and one of the chemicals they use is paraquat, which has been banned in thirty-six countries. Right on the label, it says that paraquat is fatal if inhaled."

DiPietro has been involved in the anti-GMO fight on Kauai since 2002, long before most people on the mainland had ever heard the term. Because the companies running these experimental fields are not forthcoming about their locations, or what crops they are growing, or what they are spraying on them, DiPietro had to create the maps herself. She assembled them from her explorations driving the island's dusty red back roads and looking for the tiny spray sheets the companies post alongside their fields. She has seen plenty of signs noting the chemicals being used: atrazine, lorsban, "other." (As toxic as atrazine and lorsban are, she says, it's the chemicals marked "other" that bother her the most.)

"It's supposed to be against federal law to spray lorsban in winds over ten miles per hour and to spray any restricted-use pesticides in windy conditions," DiPietro said. Here on Kauai, "it's always blowing like this."

Because the fields themselves are shielded from public view, the spray sheets are plainly not intended for the public either. They are posted to advise company workers to stay off the fields for twenty-four to forty-eight hours after a spray. This is serious business: the EPA recently announced it is considering banning chlorpyrifos, another commonly used chemical on Kauai that has sickened dozens of farmworkers in recent years, including at least ten Syngenta workers who were hospitalized in Kauai in January 2016. The workers had walked onto a cornfield twenty hours after it had been sprayed—just four hours earlier than recommended.

DiPietro had driven me by the Grand Hyatt Kauai and the Poipu Bay Golf Course, within easy drifting distance of the experimental farm. Did the golfers know what was being sprayed across the street? Would they care if they did? How about the surfers? The retirees drinking piña coladas or doing yoga on the beach? It is this lack of available information—about chemicals that are well-known health hazards being sprayed in close proximity to places where people live, work, and play—that has driven DiPietro and a host of others on Kauai to take their fight straight to the companies themselves.

A notably gentle woman, DiPietro shielded her dark hair and dark eyes beneath a baseball cap that read "Kauai Has the Right to Know." She had been to this experimental farm many times before and was not, apparently, a welcome presence. As we sat in her cab chatting, she looked in her rearview mirror and saw a white four-by-four coming up fast behind her. She sat tight. "Looks like we've got a visitor," she said.

A white pickup pulled up next to DiPietro, and a bull-necked man with fury on his face glared out from beneath a ball cap.

"Get the hell out of here and don't come back," he seethed. "And no more pictures!" The man was enraged, his voice was full of threat, and DiPietro did not try to argue. But she did not seem intimidated so much as resigned. She'd been through this ritual before. She pulled off down the road.

LAND USE ON KAUAI has a long and complex history, one that is tied up with centuries-old sugar plantations and an enormous cultural and economic gap between wealthy landowners and native and immigrant laborers. In 1920, several hundred Filipino workers staged a strike against the sugar plantations, protesting wages that amounted to less than a dollar for twelve hours of work. As they gathered, policemen climbed a nearby bluff and fired on the crowd. In what came to be known as the Hanapepe Massacre, sixteen Filipino workers were killed as they fled into a stand of banana trees. The workers were later blamed for the violence: 130 were arrested; 56 were found guilty of rioting and were imprisoned.

A few decades later, chemical companies began testing defoliants for use in the Vietnam War. "We've been a place for Monsanto to experiment for fifty years," a woman named Fern Rosenstiel told me. "They tested Agent Orange on this island right near where I was born."

King Sugar, as the industry was known, dominated the island's economy for 150 years, placing great wealth in a very few hands but also creating a plantation culture that many say remains in place today. Descendants of the sugar workers from Japan, Portugal, Polynesia, and the Philippines remain in sizable numbers throughout the state. Crippled by foreign competition, Kauai's sugar industry began to collapse in the 1980s and 1990s, and many companies picked up and left. Big Agribusiness has more than stepped into its ample footprint. The companies still hire descendants of the people who worked on the plantations—Chinese, Japanese, native Hawaiians—and these people are happy to have the work. But they also hire a lot of temporary workers from places like Malaysia.

"Their ancestors were brought here to divert rivers for the benefit of the white people who ran the pineapple plantations and the sugarcane plantations," Rosenstiel said. "Forty million gallons of water still goes out of the Waimea River through diversion, straight out into ocean, because they've never restored the diversions."

Some 14,000 acres of Kauai's land are leased to the global agrochemical conglomerates DuPont Pioneer, Dow, Syngenta, and BASF. The corporations chose Kauai because its tropical climate enables them to work their fields year-round. Company workers can plant experimental fields three seasons a year, which can cut in half the time it takes to develop a new genetically altered seed.

The "experiments" taking place on these fields consist of planting genetically engineered seeds—primarily corn—and then dousing the fields with a variety of pesticides to see which plants survive. The chemicals will kill all the weeds and some of the corn plants themselves. Between 2007 and 2012, DuPont Pioneer sprayed fields on Kauai with ninety different chemical formulations with sixty-three active ingredients, and sprayed as many as sixteen times a day, two out of every three days during the year. Statewide, Hawaii leads the nation in the number of experimental fields, with more than 1,100. Studies show that companies use seventeen times more of the highly toxic "restricted use" pesticides on experimental plots than do farmers on traditional fields.

The use of these chemicals has become necessary, at least in part, because softer, "general use" pesticides like glyphosate have begun to lose their effectiveness. Chemical companies must now engineer new seeds that will resist other, more intense chemical compounds. Dow, for example, has used its Kauai fields to develop new corn and soybean seeds that are resistant to the herbicide 2,4-D—once an active ingredient in Agent Orange that's been linked to reproductive problems and cancer.

If a corn plant can survive the chemical sprays—and if the sprays successfully kill every other plant on the field—the resistant seeds will be moved along the development pipeline; one day, this corn's progeny might end up spread across the vast cornfields of Iowa, and Nebraska, and Illinois. More than likely, the harvest from these plants will end up sweetening soft drinks or feeding the millions of cattle and pigs that supply the country's bottomless appetite for inexpensive meat.

Because GM crops have been legally declared to be the "substantial equivalent" of conventionally farmed crops, the island's farms are not required to file Environmental Impact Studies. And because of a variety of legal loopholes, including the shroud wrapped around "proprietary information," companies are not required to tell the public much of anything about what they are spraying, or where, or when.

Since they lease their land from the island's handful of large private landowners (Steve Case, the founder of AOL, owns 38,000 acres of former plantation land known as Grove Farms), the companies are largely shielded from public view. Because the companies get their spraying permits from the federal government, and not from the state or the county or the local planning boards, they do not feel obliged to answer to local complaints. And because their work is regulated by the federal government, the companies say that local laws do not apply to them. They stick to this logic even when their research takes place on thousands of acres of state land.

For the people who live on Kauai, however, the fight over GMOs and pesticides is just another chapter in a long struggle over the use and misuse of their land. They say the companies have refused to divulge what chemicals they use on their fields. They say that when people complain to the companies, they get no answers. When people complain to their elected officials, and the elected officials complain to the companies, they also get no answers. By fighting even basic disclosure laws, the companies are shutting down any possibility of understanding what consequences their chemical sprays might be having on the health of the local community. Activists, doctors, local politicians—they all want information, and they aren't getting any.

"For me, this is about the impact on our community, not on whether Doritos have GMOs or not," Gary Hooser told me. For years, Hooser, a county councilman (and thus one of the island's highest-ranking public officials), has tried to extract information from the chemical companies. He has had very little success. "I have issues with corporations controlling the food supply, but that's also not what this is about. This is about industry causing harm. I asked them politely, and in writing, for a list of the pesticides they used, and they said no, they were not going to give it to me. They were very polite."

If chemical companies on Kauai are outwardly uncooperative, their behind-the-scenes influence on the regulatory agencies charged with overseeing their work is virtually complete. Pushing states (and the federal government) to cut regulatory staff has long been a primary industry objective. Here's what this looks like in Hawaii: because of budget cuts, the state Department of Agriculture has only one employee assigned to review pesticide inspection reports. Although the department is responsible for overseeing the federal Clean Water Act, it has no statewide program for testing pesticide use in soil, air, or water. The single position on Kauai meant to monitor toxins in agricultural dust has been vacant for a year. Meanwhile, the state's health department has no programs to test for pesticide contamination in the soil, air, or water.

Kauai's sole pesticide inspector says she hasn't gotten around to reviewing most reports in several years—in part because so many concerned people have been asking her for spray records. "I've had so many requests that I haven't had a chance to work on any of my cases for so many years," she said.

As for federal oversight, the nearest EPA office is 2,000 miles away in San Francisco.

All of this means that when Gary Hooser asks companies for records about what they are spraying, he finds himself circling in an endless bureaucratic whirlpool. When he asked the state to provide a spreadsheet listing the sales of restricted-use pesticides used by Dow, DuPont Pioneer, Monsanto, BASF, and Syngenta on the island from 2002 to 2004, his request was denied. The disclosure records "are believed to contain confidential business information (CBI) or trade secrets," the state's pesticides program manager wrote Hooser. The decision made it impossible for Hooser or anyone else to determine "what chemicals are being used, by whom, at what geographical locations," Hooser said.

State law requires that companies seeking federal permits to test GMOs or experimental-use pesticides must file a copy of the request with the state. But when Hooser asked the state health department for copies of these requests, he was sent a grand total of eight.

"I said, 'There must be a problem—there must be more,'" Hooser told me. A couple of months ago, he asked again. This time, the health department said they had "a roomful of these things." We haven't even opened the boxes, the state people told Hooser, "but you're welcome to come by and look."

Although the state has an entire storeroom full of boxes, "literally nobody at the state looks at these documents," Hooser said. "Nobody. And most are highly redacted."

Companies point to reams of paper to show how regulated they are, but Hooser found that no one was checking up on them. "The state inspects them maybe five times a year, and they spray 220 days out of the year, and an average of eight to sixteen times a day. It's a tragedy. They look me in the eye and say they are inspected on a regular basis, and 43 percent of the state inspection logs are redacted."

A Hawaii Department of Agriculture (HDOA) log shows that in 2011 and 2012, the state made 175 inspections on Kauai, but more than a third of these reports had been redacted, the names of companies, employees, and alleged violations crossed out. The log has this note attached: "On two separate occasions, Kaua'i County Councilmember Hooser has requested in writing from the HDOA 'the nature of the violations and investigations without the accompanying company identification.' This information has not been provided."

When Hooser finally got his hands on a list of restricted-use pesticide sales from the state Department of Agriculture, "the core data shocked the hell out of me," he said. "Restricted use" means the chemicals (in this case including alachlor, atrazine, chlorpyrifos, methomyl, metolachlor, permethrin, and paraquat) are more dangerous—and thus more tightly regulated by the EPA—than general-use pesticides like glyphosate or 2,4-D.

"Ninety-eight percent of the restricted-use pesticides were being used by just four companies. They were using atrazine by the ton. Paraquat. Eighteen tons a year of twenty-two different kinds of restricted-use pesticides on this island only." All these chemicals didn't just disappear, Hooser knew. Some were taken up into plants, but some trickled into the island's soil, the water, the air itself.

State records show that between 2010 and 2012, the agrochemical companies purchased 13 tons (plus nearly 16,000 gallons) of restricted-use pesticides on the island. Pest control companies used an additional 74,000 pounds, mostly to kill termites and ants.

Other records show that between 2013 and 2015, companies sprayed 18 tons of restricted-use pesticides. During this period, companies also used some seventy-five different general-use pesticides, but because of lax enforcement codes, no information was available for how much was used.

Six of the seven restricted-use pesticides are suspected of being endocrine disruptors, which means they may cause sexual development defects in humans and animals, according to the EPA. Four of the seven are also suspected carcinogens. And between them, the seven have been linked to, among other things, neurological and brain problems and damage to the lungs, heart, kidneys, adrenal glands, central nervous system, muscles, spleen, and liver. And these are only the most toxic of the lot. As we have seen, even general-use pesticides like glyphosate and 2,4-D have recently been declared "probable" and "possible" human carcinogens in their own right.

A study published in March 2014 in the British journal *The Lancet* found that chlorpyrifos, a neurotoxin that is restricted in

California and many countries, is one of a dozen commonly used chemicals that "injure the developing brain" of children.

Recent hair sample testing of children living near the Kauai test fields indicated exposure to thirty-nine different pesticides, including eight restricted-use pesticides. "It's unconscionable that pesticides are being found in the hair and bodies of our children," said Malia Chun, the mother of one of the girls tested. "State and federal officials have a responsibility to ban chlorpyrifos and make sure our children are protected in our homes and schools from these hazardous chemicals."

But it wasn't just chlorpyrifos. Children were exposed to "a cocktail of pesticides, and the consequences of exposure to such mixtures over a lifetime are not known, nor is the issue of exposure to such mixtures currently evaluated by our regulatory agencies," said Emily Marquez, an endocrinologist and staff scientist at the Pesticide Action Network.

Also in the cocktail: permethrin, a suspected carcinogen thought to compromise kidney, liver, reproductive, and neurological function. When combined in the body with chlorpyrifos, permethrin has been shown to be "even more acutely toxic," according to E. G. Vallianatos, a twenty-five-year veteran of the EPA and author of *Poison Spring: The Secret History of Pollution and the EPA*.

Another ingredient in the cocktail: atrazine, the second most widely used herbicide (behind glyphosate) in the United States. A known carcinogen, atrazine is sprayed on half of all corn crops and 90 percent of sugar sold in the United States—which is why it is commonly used on experimental fields in Kauai. "A little bit of poison to an adult is a lot of poison to a developing baby," Dr. Tyrone Hayes, an endocrinologist at the University of California, Berkeley, told an audience on Kauai recently. The poisoning of a young child can cause health problems that can last a lifetime,

Hayes said; his own research has found that frogs exposed to barely detectable levels of atrazine developed both male and female genitalia.

On Kauai, frustration with chemical company behavior grew most acute in the town of Waimea, on the island's west side. In 2000, residents of the town filed a formal complaint claiming that pesticide-laden dust was blowing into their homes from experimental fields operated by DuPont Pioneer. They got nowhere.

Six years later, sixty students in a Waimea school went to their health office complaining that a "chemical smell" was making them nauseous and dizzy. Some students fainted. Others were seen covering their noses with their T-shirts. Nearly three dozen were sent home. A local reporter noted that several of the children "had their heads in their hands and tears in their eyes."

The school is situated just a few dozen yards from experimental fields leased by Syngenta. Firefighters, police, a hazmat team, and officials from the state health and agriculture departments descended on the school to examine students and take samples from the nearby fields.

At first, company and state officials blamed the outbreak on a malodorous plant called *Cleome gynandra*, also known as wild spider flower or (more accurately) stinkweed. "It does stink and as a company we certainly hope the children are feeling better," a Syngenta official said.

Though it is eaten (boiled) in some parts of the world, stinkweed has been known to cause headaches and even nausea in some people who are particularly sensitive to it. But Gary Hooser, who was a state senator at the time, was not convinced. He started making phone calls. He wanted the company, or the state, to tell parents what chemicals were being applied to the crops near their children's school. Neither state officials nor Syngenta would tell the senator anything, and repeated attempts by local reporters "to compel authorities to release the information were unsuccessful."

Company claims about stinkweed contamination struck some scientists and doctors as disingenuous. Given that the company fields were so close to the schools and to local homes, a few things were beyond dispute. There was no questioning the presence of restricted-use pesticides, or that dust from these pesticides routinely migrates into residential properties, or that the chemicals have a well-documented connection to childhood neurological problems, including autism, ADHD, and fetal brain defects, wrote J. Milton Clark, a professor at the University of Illinois School of Public Health and a former senior health and science adviser to the EPA, who examined the evidence for an island task force on pesticides.

There was no evidence to support the stinkweed theory, Clark wrote. "Symptoms of dizziness, headaches, nausea, vomiting, and respiratory discomfort are consistent with exposure to airborne pesticides," he wrote. The children's symptoms "were far more likely related to pesticide exposures than from exposure to stinkweed." If the companies continued spraying, Clark recommended that local health centers near agricultural fields be given kits "to quickly test for organophosphate poisoning."

It took nearly six years for state health officials to formally weigh in on the incident. When researchers from the University of Hawaii sampled the air around the Waimea Canyon Middle School, they indeed found evidence of stinkweed. But they also found five pesticides, including chlorpyrifos, metolachlor, bifenthrin, benzene hexachlorides (BHCs), and even DDT, which has been banned in the United States for four decades. Although the chemicals were found in amounts below EPA health standards, the presence of agricultural chemicals was clear evidence of "pesticide drift," according to Hawaii's Department of Agriculture. How many years these chemicals—and perhaps dozens of others—had been drifting into Waimea homes and schools was not addressed.

To Gary Hooser, Waimea's pesticide drift was just part of the problem. The larger issue was the way companies seemed to consider themselves beyond the reach of public oversight. "The failure to release the information about what is sprayed out there only increases the public's mistrust that something harmful is being sprayed," Hooser said at the time. "They know what was sprayed out there and they should tell the public."

What the island needed—and what the medical community began demanding—was information about the chemicals being sprayed in their communities. Frustrated by the lack of quantitative data about pesticide use, a group of west-side physicians wrote that they had "many qualitative examples that point to a higher than normal incidence of many ailments and disease processes occurring in our patient populations." They'd seen birth defects involving malformations of the heart that were occurring at ten times the national rate. Miscarriages, gout, cancer, hormonal imbalances—all were occurring at unusually high levels, the doctors wrote, noting that Hawaii had not had surveillance for birth defects since 2005. They called for epidemiology studies by the CDC and Hawaii's Department of Health to better understand the causes.

"We all share a deep concern for the health of our patients and the concern of what may be happening to our community by being exposed to this unique cocktail of experimental and restricted-use pesticides on an almost daily basis," the Kauai doctors wrote. "We need to understand what chemical toxins are being sprayed, how often they are being sprayed, and how close our patients live to the specific areas being tested with these pesticides. It is unconscionable to allow open-air testing of new combinations and untested chemicals in any location that cannot guarantee the separation of the testing and any unwilling or unknown exposure potential to the public."

The doctors' worries reflected conclusions in a major study by the American Academy of Pediatrics (AAP), which contended that a growing body of evidence points to associations between pesticide spray exposure in young children and a range of diseases, from childhood cancers to autism. On Kauai this was especially worrisome for the children of people who work in the fields, said Dr. Lee Evslin, a pediatrician on the island. The AAP "never had a mandate about pesticides before, but they have now placed it in our laps," Evslin said. "This body carries a lot of weight, and they are basically saying to the pediatricians of the world, 'pay attention to this. These are dangerous substances.'"

Margie Maupin, a nurse practitioner on the island's west side, said the presence of so many pesticides—and so little information—had left her unable to do her job properly. "Thousands of reputable studies have already been done that show pesticides are known hazardous toxins," she said. "The probability that these pesticides will hurt a lot of people on the west side, I believe, is high. Some health care providers are already seeing signs of serious illness and disability now, and we are at a loss for how best to protect our patients from this onslaught of known, dangerous exposure."

Taking the Companies to Court

When I visited Waimea, I met a man named Klayton Kubo, who has been raging about clouds of dust for fifteen years. When we first sat down at a picnic table in the town center, Kubo refused to talk to me. Too many people around, he said, looking over his shoulder. The companies know who I am.

Instead, we drove to the top of a nearby ridge, parked, and walked along a dry path overlooking the town. To our left, in the near distance, we could see fields operated by both DuPont Pioneer and Dow. Tractors were working the fields, with red dust rising behind them. Perhaps six miles away, the largest of the plumes rose hundreds of feet into the air.

"If you think this is bad, you should come back during a trade winds day," Kubo said. "It's fucking insane!

"Two hundred yards outside my living room window, I can see their facility. The wind comes this way, we get it. The wind goes the other way, we get it. And right in the middle is a school and a town."

Kubo pointed at the plume in the distance. "What you see right there? That's what's in my kitchen," he said. "I scrape the stuff off my glass-top stove. That's why I've been grumbling the longest."

As we walked down the hill, an official-looking white pickup truck drove by. "Ha! Syngenta!" Kubo shouted. "Don't fuck with my truck!"

In 2011, more than a hundred of Klayton Kubo's neighbors filed a lawsuit against DuPont Pioneer claiming that dust from the company's fields was damaging their property. Despite more than a decade of complaints and a formal citizen petition seeking relief from pesticide-laden dust, the lawsuit claimed, Pioneer's GMO operations continually generated "excessive fugitive dust" and used dangerous pesticides "without taking preventative steps to control airborne pollutants as promised by Pioneer and as required by state and county law."

"The community is covered," the plaintiffs' lawyer Gerard Jervis said. Residents are "living in lockdown, unable to open their doors or windows." The suit pointedly did not make any health claims, though Jervis said local residents complained frequently of asthma and chronic obstructive pulmonary disease.

A company spokesperson defended Pioneer's practices. "We operate our facilities on the islands with the highest standard of safety and environmental responsibility and we plan to vigorously defend our case."

At the beginning of the trial, when residents alluded to health problems they attributed to the dust, the judge in the case reminded his attorneys that the case was about property damage only. The case was not about the effects the chemicals might be having on their health.

Jervis reminded the court that the EPA requires that applicators must not allow spray to drift from fields into private property, parks and recreation areas, woodlands, or pastures. He also noted that the state's air quality study did not even try to look for more than thirty pesticides that have been used at the GMO test fields since 2007, including two of the most heavily used and dangerous: methomyl, an insecticide, and paraquat, a weedkiller that (like atrazine) has been linked to Parkinson's disease and (also like atrazine) is made by Syngenta. Paraquat has been banned both in Switzerland, Syngenta's home, and across Europe.

As the Waimea lawsuit proceeded through its paces, worries about pesticides on Kauai continued to grow. A local Kauai diver discovered a massive die-off of up to 50,000 sea urchins. A biologist for the state Department of Land and Natural Resources speculated that the chemicals sprayed on GM seeds might have been a cause, because when it rains, the loosened red topsoil on treated land flows into streams and rivers that eventually flow out into the ocean and onto coral reefs.

"Kaua'i produces more GMO seeds than anyplace," Don Heacock, the biologist, said. "Now, there are a whole bunch of people in the genetic engineering camp that say GMO crops need less pesticides, but the new wave of crops is more toxic than ever before. The

Bt corn is meant to kill. It has an insecticide protein in the corn. In the Midwest, they found that the residue from GMO corn is related to aquatic insect deaths, which are food for baby fish."

That same winter, the internationally renowned environmentalist Vandana Shiva traveled from New Delhi to Kauai to speak to anti-GMO activists. "I think your island is truth-speaking to the world that GMOs are an extension of pesticides, not a substitute or alternative to it," she said. "[Hawaii] has become like a nerve center for the expansion of destruction. GMOs are not a safe alternative to poisons, they are pushed by a poison industry to both increase the sale of the poisons and simultaneously monopolize the seed."

Evoking the 1984 disaster in Bhopal, India, when a chemical leak from a Union Carbide plant (now a subsidiary of Dow Chemical) killed and injured tens of thousands of people, Shiva said that chemical manufacturers had long since transformed themselves into the biotech industry. "War and agriculture came together when the chemicals that were produced for warfare lost their market—and the industry organized itself to sell those chemicals as agrochemicals," Shiva said.

Energized, activists on Kauai decided to take their animus against the companies to the streets. In December 2012, Fern Rosenstiel, who grew up near Agent Orange test fields, organized a small protest by the Kauai airport. She was joined by Dustin Barca, a professional surfer who, at the age of twenty-six, had become a successful professional fighter in mixed martial arts. Surfing and fighting had made Barca famous on Kauai and around the state, and he decided to leverage his fame to galvanize people against the chemical companies.

Barca had an idea. That same month, during the Pipeline Masters surfing competition on Oahu, he made headlines just by standing on the beach.

"There were 30,000 people on the beach, millions more [watching] on TV," Barca told me. "Me and this little kid carried around a bright red and yellow banner that said 'Monsanto's GMO Food Poisons Families.' That was my first, initial move to get the word out, on the north shore of Oahu, the most famous surf spot in world."

When I met Barca, he, like Klayton Kubo, refused to talk in public. He didn't know who might be watching. But even more than Kubo, Barca is used to fighting. He has the wiry frame of a welterweight. He is missing teeth. His ears have been so damaged they have turned inside out. Ever since he'd entered the political fray, he's had people videotaping him, he said. At a recent anti-GMO rally, he confronted a man taping him with a video camera. "I told him, 'Whoever sent you is going to have to do better than that,'" Barca said.

"Are these companies good for people or nature? How can we tell if they don't give us the information?" Barca said. "We know what they're doing. They've admitted they're spraying 2,4-D near our communities, and the trade winds blow every single day. We've gone so far into a place where everything is done behind closed doors. It was the same thing that Dole and others did to overthrow the queen. History repeats itself. You just have to know the blueprint to catch it."

Emboldened by the anti-GMO energy he felt at the surfing tournament, Barca decided to see just how much energy he could leverage across the state. He and Rosenstiel set about organizing marches on all five islands where companies were testing GMOs and pesticides.

On Oahu, close to 3,000 people turned out for a rally in the pouring rain. "I thought, 'Holy shit, that's a lot of people who feel like I feel,'" Barca said. "That set up the momentum.

"We went to a different island every Saturday. First was Honolulu. There's a town there that is the Waimea of Oahu, surrounded by Monsanto experimental fields. We went to a high school over there. You literally walk fifty feet behind their fields. All the kids are running around, these are experimental fields. There are giant aerial sprayers. They can spray 250 times a year, dozens of times a day."

As word spread, the anti-GMO crowds continued to turn out in droves: 300 came out on Molokai, one of the smallest of the islands; 1,500 on the Big Island; 2,000 on Maui.

Meanwhile on Kauai, with anti-GMO energy reaching a peak, Gary Hooser found himself in a bind. If he encouraged activists to stand out in front of expensive tourist hotels, holding up signs saying that Kauai is "Ground Zero for Experimental GMOs," his community stood to lose tourism dollars. He decided instead to introduce a bill that would force companies to do what they so far had refused to do: disclose what they were spraying, on what crops, and in what fields, "to see if we have anything to be afraid of." The bill also sought to create no-spray buffer zones around schools, homes, and hospitals. His bill carried criminal sanctions for companies that refused to comply; Hooser hoped this would at the very least encourage whistleblowers.

"People were concerned with pesticides and GMOs, so what was I supposed to do?" Hooser told me. "I met with the companies, asked them to give me their data, asked them to help me separate the wheat from the chaff, and the companies wouldn't tell me anything. They wouldn't respond to my questions. They lied to me. They were telling me they 'only use what other farmers use.' No other farmers use this stuff, and not in anything like the toxicity or the volume. The more they lied, the more I dug into it, and the more angry I got."

Industry executives claimed the bill's disclosure rules were unnecessary, unfair, and pseudoscientific. Alicia Maluafiti, the executive director of the Hawaii Crop Improvement Association, a bio-tech trade group, called Kauai's move "a pretty pissy bill."

"It's not about community health, it's not about pesticide use, it's about getting rid of these companies," she said. She called the pesticide disclosure bill "fearmongering by Mr. Hooser and the extremists on Kauai."

Companies dismissed complaints by repeating that both GMOs and pesticides were highly regulated by the government. Genetically engineered products "have been out there for seventeen years now," said Mark Phillipson, Syngenta's head of corporate affairs in Hawaii. "There have been 3 trillion meals served that have had genetic-engineered components in them, and not one reported incident, acutely or long term, associated with GM causing an allergen or toxicity issue."

During the hearings on the bill, hundreds of people from both sides showed up to voice their opinions, many of them wearing colored shirts to show which side they were on.

"We made shirts with red and yellow, representing the strong in Hawaiian tradition. They wore blue," Dustin Barca told me. "It was almost like the Bloods and the Crips."

Companies urged their employees to show up en masse to counterbalance the protesters. "The companies bussed workers in here so we couldn't even get in to testify," Rosenstiel said.

Indeed, the battle caused a lot of collateral damage in the Kauai community. "We had a number of doctors come forward—a clear majority of pediatricians signed a letter supporting the bill—but even they paid a political price," Hooser told me. "These doctors get hammered. They didn't say they 'know illnesses are caused by this spraying,' they just said they were concerned. But the pushback by the companies, their bloggers, the media stuff, it's been intense."

During one hearing, a councilman asked an official from the

state Department of Agriculture if there was any evidence of pesticide drift. Complaints do come in, the official said, and the state goes to houses, swipes the windows, and sends the samples out for testing. When the investigation is complete, the neighborhood is notified. The whole process—if it actually gets completed—can take two years.

What if it's a pregnant woman or a child who's being exposed? Gary Hooser wanted to know. What good is a two-year lag in the testing to them?

As the vote neared, Rosenstiel and Barca helped organize another march. Some 4,000 people marched to the Kauai County Building to support the bill. Some wore gas masks. Others wore death masks. Many wore red T-shirts with yellow letters saying "Pass the Bill."

Finally, after a hearing on the bill that went on for nineteen hours straight, the Kauai County Council passed Hooser's bill, 6–1. The mayor vetoed the bill, but the council overrode his veto. It was official: Kauai's anti-GMO activists had pushed their elected officials to pass a bill requiring some of the world's most powerful companies to disclose what pesticides they were spraying and where. In a very real sense, the vote was a watershed.

Yet within weeks, DuPont Pioneer, Syngenta, BASF, and Agrigenetics Inc. (a company affiliated with Dow AgroSciences) sued the county in federal court. Their argument: Company farming practices adhere to state and federal laws. Local laws have no jurisdiction over them.

In August 2014, federal judge Barry Kurren agreed with the companies that the state pesticide law preempted any county law regulating pesticides.

An attorney representing two of the companies said she was very pleased. "This is what we told the county when they were discussing it initially," she said. "I think they wasted time, effort, and money trying to fight for a law they had no right to pass in the first place."

Gary Hooser saw the ruling differently. "We passed the bill with a democratic process, with thousands of citizens involved," he told me. "We got the votes like we were supposed to. We overrode the mayor's veto. And they sued us for the right to spray poisons next to schools." The anti-GMO forces on Kauai have appealed the judge's decision; it is now awaiting a hearing in federal court.

Before the dust from the political fight could settle, Dustin Barca, the surfer and professional MMA fighter who had done so much to organize the anti-GMO rallies, decided to make one last public push: he ran to unseat the mayor who had vetoed Hooser's bill. During the campaign, he ran—literally, ran—around the island; three marathons, back-to-back. Although he didn't win, he did pull 40 percent of the vote.

"This was totally untypical of me," Barca told me. "I just had a voice in my heart and my head that said, 'You have to do something about this right now.' I threw my whole selfish life away and went into selfless life. I'm not doing this to get rich or famous. I could be making millions fighting in the UFC [Ultimate Fighting Championship]. I'm here for my kids. No other reason."

About this time, the state Department of Agriculture and Kauai County agreed to set up a fact-finding effort to look into pesticide use. They recruited nine volunteers with backgrounds in agriculture, environmental health, and toxicology. Kauai County split the $100,000 cost of the study with the state Department of Agriculture.

"The big question, the meta-question if you will, is: Are people being harmed from pesticides being sprayed by GMO companies?" said Peter Adler, a veteran mediator who will oversee the project. "We hope to really present some pretty rigorous inventories of what we know, what we don't know, and what we need to know still and find out. People are talking at their conclusion levels and we want to get down to: What's the data? What's the evidence?"

For local residents, there were other "meta-questions," like whether they should have a say in how their land is used, and how they can protect their own neighborhoods. They have had some victories: in May 2015, a federal court jury awarded $500,000 to fifteen Waimea residents who claimed the red dust from DuPont Pioneer fields caused "loss of use and enjoyment of property." The verdict said that DuPont Pioneer failed to follow generally accepted agricultural and management practices from 2009 to 2011; the jurors found the "seriousness of the harm to each plaintiff outweighs the public benefit of Pioneer's farming operation."

Ten days after the verdict, DuPont Pioneer shut down its 3,000-acre experimental field operation in Kekaha. It plans to consolidate it with operations on Oahu.

At the end of April 2015, Gary Hooser flew to Switzerland to speak at a Syngenta shareholders meeting in Basel. He wanted to ask the company to stop using chemicals in his district that are already illegal in the company's own country—indeed, across the company's own continent.

The company did not welcome him. On his blog, Hooser recently wrote:

> Syngenta did not want me there and was working on many levels to prevent me from speaking, but legally there was nothing they could do to stop me . . .
>
> I asked them to withdraw from their lawsuit against the County of Kauai, to honor and follow our laws, and to give our community the same respect and protections afforded to the

people in their home country of Switzerland. I pointed out that their company uses highly toxic Restricted Use Pesticides (RUPs) in our community, including atrazine, paraquat and four others that they are forbidden by law from using in their own country.

We are not going away and we will not tap out. So long as these companies continue to disrespect and disregard the wishes of our community, we will continue the battle to make them comply.

Fern Rosenstiel, who had organized so many of the marches on Kauai and the other islands, accompanied Hooser on his trip to Switzerland.

"For me, this island is the trunk of the tree," Rosenstiel told me. "If we can get these companies off this island, if we can cut this tree down, it will cause a positive worldwide reaction. I'll be here until the day I die, or until these guys are gone."

6.

Fighting for That Which Feeds Us

Around the time Kauai voters were rattling the biotech world by approving a pesticide disclosure law, a group of indigenous Hawaiians and back-to-the-land farmers on two other Hawaiian islands—Maui and the Big Island—were going a dramatic step further: they were pushing to ban GMOs altogether.

To the big agrochemical companies, this was a far more dangerous game. Being forced to tell people what they were spraying on experimental farms was one thing. Being voted off an island—by what amounted to a pair of tiny county ordinances—was something else entirely.

The Big Island, basically, had one GM crop—Dennis Gonsalves's papaya—and wanted to lock the door tight before any of the big companies moved in. Maui was a different story. To companies like Monsanto, Maui was not just a warm place to test out new crops; it was the very center of global GM seed production. The majority of the corn seed Monsanto sells to farmers in its biggest markets—Argentina, Brazil, and the United States—originates on Maui. If the island's voters got their way, Monsanto and Dow AgroSciences (the other biotech giant operating there) would have their GMO operations shaken at their foundation.

Beyond this, of course, was the ongoing global perception game. It was one thing for companies to lose fights in Europe—GMOs had never been welcome there—but losing another major public relations battle in the United States was something else altogether. Banning GMOs on a couple of little islands could ignite larger movements in bigger places that were already primed for the fight. Vermont. Oregon. California. And then? South America? India?

For the companies, already shaken by the Kauai vote, the battles on Maui and the Big Island were about global markets and their ambition to sell seeds and chemicals to the world. They would spend millions of dollars to prevent the anti-GMO ball from rolling any further. There was no way they were going to let a small group of activists derail their global business plans.

But for the people on the islands themselves, the battles were far more intimate. To them, the fight against GMOs resembled similar fights not in the United States but in the developing world, where indigenous people and political activists had struggled against global conglomerates for years. They were fighting to protect land they considered sacred. They were fighting to break a long history of colonial oppression. They were fighting for the right to feed themselves.

The Battle on the Big Island

Even as Dennis Gonsalves traveled the world trying to persuade farmers to adopt his beloved papaya, his neighbors back home on the Big Island were working just as hard trying to ban GMOs altogether. In a way, the anti-GMO activists took the same line as

Dennis Gonsalves: they wanted to protect farmers. It's just that the farmers they wanted to protect were of an entirely different sort.

Nancy Redfeather is not particularly interested in whether GM papayas continue to sell in China or anywhere else. She wants her island to grow food for itself. All this technology, all these companies, all this talk of a globalized food economy—it all just gets in the way of growing nutritious food for people who live down the road.

The day I met her, on a stunning 70-degree day, Nancy poured me a glass of tangelo juice her husband, Gerry Herbert, had made from one of the thirty-six varieties of fruit trees the couple grow on their one-acre organic farm. Nancy offered me a cup of coffee, ground from beans they roasted from the twelve varieties of coffee they grow at home. She offered me a plate of fruit—apple bananas, yellow dragon fruit, navel oranges, blush pink grapefruit, star fruit, Tahitian pamplemousse, avocado—all just picked from their farm. Had I stuck around for dinner, we might have eaten a meal made from kabocha pumpkin with cloves, turmeric, ginger, and garlic. Plus wild chickens or wild pigs. (In six months, Gerry caught thirty-nine feral pigs in a trap. Their meat is exquisite, he says; given their proximity to his crops, the pigs eat better than most people.)

Nancy and Gerry run a small organic farm near Kona. Nancy moved to Hawaii from California in the mid-1970s, when the back-to-the-land movement sent many mainlanders looking for places to set up sustainable livelihoods. They built their timber-frame home themselves. They have a kitchen inside the house, and another one outside the house. Three-quarters of the food they eat they grow themselves.

After lunch, Gerry gave me a tour of his gardens. Here is a sample of what he grows in a single acre: There were trees called jaboticaba (Tupi for "fat of the flesh of the turtle") that had strange black berries growing straight from the bark. The berries resemble hefty Concord grapes and yield beautiful pink juice. Gerry freezes this juice, then uses it to make banana bread.

There were four varieties of black beans, lychee trees, a Rajapuri banana tree that produces 500 pounds of fruit a year. There was an eighty-seven-year-old mango tree that still drops 250 pounds of fruit a year ("We eat as much as we can and feed the rest to the chickens," Gerry said). Black-capped raspberries. Star fruit. Pigeon peas. Dragon fruit growing along a stone wall; coffee bushes that produce 1,500 pounds of beans a year; 120 pounds of macadamia nuts. Five different kinds of avocados, including 180 pounds from a single tree. The Yama avocado, Gerry says, makes Hass avocados "seem like something you'd only feed to the pigs."

"I've lived all over the States and all over the world, and this is the best growing climate I've ever lived in," Gerry said.

Gerry got his agricultural degree from UC-Davis, near where the Flavr Savr tomato was first developed, and then spent thirty years farming twenty-two acres in Mendocino. When I asked Gerry if he had ever tried a Flavr Savr, he smiled.

"I tasted the Flavr Savr. It tasted like rubber," he said. "I thought, 'Wow, you guys are never going to sell this,' and sure enough, it fell on its face.

"If corporations develop a plant, they develop it for their own reasons. They don't develop it for nutrition. They could care less about nutrition. That's not the people you want growing your food."

Especially given Hawaii's utopian weather and soil, Nancy and Gerry think that using the state's precious land to grow GMOs—including Dennis Gonsalves's papaya—is a travesty, and symptomatic of a farm system focused entirely on making money for exporters. Hawaiian farmers could provide close to 40 percent of the state's fruits, but rather than sell them locally, companies ship them to the mainland. "We keep one percent," Nancy said. "You can't even find it in stores. By the time it gets somewhere else, it loses its taste and its nutrition—just like the food we import."

She pointed to my plate, brimming over with fresh-picked organic produce.

"Nothing on that plate can you find in stores," she said.

Nancy and Gerry's farm is typical of how most farming is done on the Big Island: 80 percent of the farms are under five acres. Their farm creates virtually no waste; the couple generates 1,500 pounds a year and puts all of it back into their soil. "You can think of this place as a mini experimental station for home producers," Nancy told me. "It's intended to be that. We don't just grow what we know we can grow. We try all kinds of things. We have a lot of failures and a lot of successes. We're also trying to be sustainable, trying to grow with only inputs from right here on the farm."

Except for GM papaya, the only biotech crop grown on the Big Island is a few hundred acres of corn, grown to feed cattle. None of the big companies have tried to push their experimental corn and soy operations. Yet.

"We are a land of small farms," Nancy said. "The biotech industry didn't come here. We don't have big, flat land they want to grow crops on. It's not as good for them."

Imagine if the state reorganized its priorities and started buying food from its own farmers, Nancy said. Imagine if it started providing local schoolchildren with fresh produce from right here on the islands, rather than processed food from the mainland? Hawaii spends $470 million a year on obesity care and hardly anything on prevention—and it is imported, processed food that is making people fat. And consider this: virtually all of Hawaii's food imports come through the ports of Los Angeles and San Francisco. If those boats stopped coming here—if there was an earthquake or a terror attack—"Hawaii would have one week before people started to starve."

"When you put chemicals into the ground, it wipes out all the critters—the fungus, the bacteria—that produce fertility," Gerry said. "We had a hundred years of sugar, and now the soil is just loaded with toxins: lead, arsenic, DDE, Agent Orange. It's just loaded. Now all your nitrogen producers are dead, and you have to buy synthetic fertilizer. It's like an addiction, and after a few years, the land is burned out. The soil is dead. It's a red powder. Even weeds won't grow there."

Back in 2000, as the local GMO debate began to heat up, Nancy did some research and discovered there were 4,000 experimental field trials going on all over the state. What the companies were growing, and what they were spraying, was a complete mystery. "No matter who you asked, no one knew what this was," Nancy told me. "The Big Five companies were all here. So we—five mothers of young children—started looking into it, and decided the community needed to know what was happening here."

Nancy turned to politics and found an ally in Margaret Wille, a Hawaii County Council member. Wille is as adamant about protecting farmers as Dennis Gonsalves and Nancy Redfeather, but when it comes to GMOs, she falls squarely into Redfeather's camp.

Especially given volcanic debates about GMOs brewing on Kauai and Maui, Wille considered herself a bulwark against industrial agriculture on her own island. "We look around and see what's going on in other counties," she said. "On Maui, a major section of agricultural land has been taken by these GMO corporations. Now we have dust storms because most GM corn is done with herbicide-resistant chemicals, which kills the soil, makes it sterile, and makes it unstable, so you get dust storms.

"My district is a breadbasket district. A lot of it is organic, and there is a whole culture of protecting the land, of planting indigenous crops and heirloom seeds. I've heard GMO people say, 'We're going to be everywhere so you won't have any choice.' It's like having an invasive species or a virus: you can't protect against it. As a culture—we have a big native Hawaiian population—we're going in the opposite direction."

In a move that made international headlines, Wille introduced a bill in 2013 that would ban GMOs from being planted on the Big Island. Papaya plantations (and corn silage farms) would be grandfathered in, so there was no risk that Dennis Gonsalves's brainchild was in any danger. But no other land would be available to industrial, experimental farms. Wille wrote her bill "to prevent the transfer and uncontrolled spread of genetically engineered organisms on to private property, public lands and waterways." But the larger question was clear: voters on Hawaii should have a say in how their land is used and by whom.

After a great deal of rancorous debate in the county council, the bill was approved.

Big agricultural companies—worried that the decision would serve as another domino in the global anti-GMO movement—immediately sued to prevent the county from enforcing the law. Lawyers representing major industries—the Hawaii Floriculture and Nursery Association, the Hawaii Papaya Industry Association, the Big Island Banana Growers Association, and the Biotechnology Industry Organization, the world's largest biotech trade association—claimed the bill lacked scientific evidence. In the two

decades since Dennis Gonsalves began his work, genetically modified farming had become a "critical and generally accepted part of
agriculture," their complaint said.

Industry also claimed that Wille's law was invalid, since local
ordinances don't trump state or federal law, and in November 2014,
U.S. Magistrate Judge Barry Kurren, once again, agreed: county
law could not override state and federal law. The law banning
GMOs was overturned: the county is appealing in federal court.

Industry representatives were elated. "This is something to be
thankful for," one of the plaintiffs' attorneys said. "This is really
important to some of the farmers. It has a big impact on their lives
and their livelihoods."

Nancy Redfeather scoffed at such statements.

"The Big Ag industry says, 'We're a $270 million industry,'" she
said. "We say, 'What are your products? What do you sell here?'
The answer is: 'Nothing.'

"We want to be like Vancouver Island: lots of local organic
farms," Redfeather said. "That's what I want: to create jobs, healthy
food, more dollars floating through our own economy. That's what
a lot of people were thinking when we passed this bill. It was really
arrogant of Judge Kurren to say, 'It's not the responsibility of the
county to regulate what they want.' That 'the health of the land is
none of your business, it's the business of the state.' When you look
at the state budget for the Department of Ag, the appropriation for
local agriculture is so small you can't even see it on a bar graph. The
state is not capable of protecting us from anything."

But industry didn't stop there. Given their success in court, companies turned their attention to unseating Margaret Wille, their
nemesis on the county council.

"The super PACs all lined up against me," Wille told me. "They
spent hundreds of thousands of dollars to defeat me. They sent out

fliers with a papaya on it, saying, 'A vote for Margaret is a vote against the community.' They flooded mailboxes with massive mailings, they did phone calls, they went door-to-door leaving all kinds of negative stuff. They brought people in from Honolulu. It was really the first time that big money came in to defeat a local councilperson."

After a "tsunami of outrage and objection," Wille survived the onslaught. "We are tired of having these lobbyists control things," Wille told me. "The fact I can win against tremendous odds and money and manpower is very hopeful. This isn't over yet."

The War on Maui

Indeed it wasn't. If anything, the skirmish on the Big Island was just a preview for the real fight, which was already under way across the water on Maui. This time, the global seed companies were not going to wait around for a vote to turn against them. They couldn't afford to: their experimental fields on Maui were the very heart of their global GM seed business. A loss on the Big Island, where there were no experimental farms, was largely symbolic. A loss on Maui would be catastrophic. On Maui, the companies would have to use their muscle—more than $8 million worth—to convince island voters that GMOs were good for them.

Before agreeing to meet with me, Alika Atay had to consult the moon. He checked his calendar. I'd be arriving on Maui in late March, during a new moon phase. He'd be planting, he said, but could meet me late in the afternoon.

On my way to meet Alika, I stopped in a local Safeway supermarket to see what kind of fruit was for sale on an island that can produce virtually anything. What I found was the same fruit you would find in a Giant in Baltimore or a Kroger in Dallas or a Piggly Wiggly in Atlanta: Bananas from Costa Rica. Apples from New Zealand. Oranges from Florida.

As far as I could tell, it was pretty much impossible to buy fruit grown down the street.

For Alika, as for Nancy Redfeather, this is precisely the problem. In their eyes, the fight against GMOs is part of the much larger fight to loosen the stranglehold that large food companies have on their beloved local food economy. Despite unparalleled weather and growing conditions, the share of produce the state grows for itself has fallen by half since 1990; it now imports two-thirds of its fresh fruits and vegetables. In 2009, for the first time, Hawaii had more land planted for experimental seed crops than for growing fruits and vegetables.

Hawaii's agricultural experts have estimated that replacing just 10 percent of the island's food imports with locally grown produce would create 2,300 jobs and $313 million in the local economy and generate nearly $200 million more in sales and tax revenues.

"The state Department of Education serves 50,000 meals a day, and 90 percent of the food comes from imports," Alika said. "I went over there once and asked to see their order sheet. The first two items on the list were five million pounds of apples and five million pounds of oranges.

"I said, 'You guys are part of the fucking problem! You say you want to be sustainable, and then you order ten million pounds of apples and oranges from the mainland? Why not order ten million pounds of tangerines and guava and papaya and star fruit that we grow right here?'"

The day we met, Alika, as he is known, was dressed in jeans and a green "MauiThing" T-shirt adorned with a pitchfork. A camouflage baseball cap barely contained the white curly mane that cascaded down his leathered face and neck. As we talked, Alika's cell phone continued to ring; fellow farmers were checking in about two issues on which Alika has become a charismatic leader: farming and politics.

Alika is the president of the Hawaiian Indigenous Natural Farming Association and a leader in the anti-GMO group called the SHAKA Movement, named for the local hand gesture (a fist with thumb and pinkie extended) used to express cultural solidarity. He is both a grower (he plants, among other things, cucumbers, tomatoes, several varieties of sweet potatoes, and apple bananas) and an educator. He spends a great deal of time teaching sustainable practices to young farmers. He wants them to learn about "canoe plants," the crops that Hawaii's original settlers brought over in tiny boats as they crisscrossed the islands of the Pacific.

"Our ancestors were pretty cool," Alika says. "Generations ago, they selected particular plants, and for 1,700 years they survived and thrived. They fed millions. And their farming was 100 percent organic. *Nothing* was imported.

"Now, we're being asked to grow European seeds, and our soil doesn't have the same geologic composition as it does in the Northeast or in Europe. The cattle and pigs raised here eat our crops, then get 'finished' on the mainland, where they shit out our minerals on someone else's land."

Alika sees the struggle against industrial agriculture as far more than just trying to rid his island of GMOs, or pesticides, or global conglomerates: it's about preserving *aina*, the Hawaiian term for "that which feeds us." *Aina* represents a sacred bond between people and a place that, once broken, threatens to destroy both humans and the world around them. In their fight against GMOs, Alika and the SHAKA movement considered themselves, as their ancestors did, to be "*aina* warriors."

Forty or fifty years ago, the pineapple plantations sprayed DDT and it leached through the soil, reached local aquifers, and contaminated drinking water. Forty years later, they went back and tested it, and the same wells were *still* contaminated with DDT. Then, in the 1980s and 1990s, heptachlor was being sprayed on the pineapples. The plants got cut up and fed to cattle as "green chop." Then the milk was bottled and served to local kids.

"For us, this is intergenerational oppression," Alika said. "It's the mentality of the plantation, but instead of plantation bosses, now it's biotech corn bosses. How can you convince people to free themselves from the bonds of oppression?

"There are two types of power: organized money and organized people," he said. "With organized money, you see the long arm of corruption. They can pervade all levels of government. People make all kinds of decisions because of power and money. When you hear them say, 'We're here to feed the world,' they forgot one word: 'Well.'"

IT IS NOT JUST NATIVE HAWAIIANS who revere the Maui landscape and are willing to fight to preserve it. As on the Big Island, Maui has also been a magnet for back-to-the-land white farmers from the mainland who share the native resistance to corporate agriculture.

Gerry Ross and his wife, Janet Simpson, moved to Maui in the 1990s to take over her parents' farm in the middle of the island. Janet left a career as a coffee roaster outside Calgary; Gerry quit his job as a PhD geologist who worked in the Arctic for the Geological Survey of Canada. Today, Gerry is a trim man with a pair of studs in his left ear and two rattails dangling from beneath a dirty white baseball cap; his organic farm produces potatoes, sweet potatoes, lettuces, kale, broccoli, beets, carrots—"anything you could possibly want to eat except strawberries or asparagus."

It was not always thus. When Gerry and Janet first took over the farm twenty years ago, the local agricultural extension agent told them the first thing they needed to do was fumigate the soil with fungicides and atrazine.

"My father-in-law passed away fourteen months after we got here, from cancer," Gerry Ross said. "The guy at the ER asked what he did for a living. I said, 'Farmer.' He said, 'Yep. We see it all the time.' I'd be willing to bet that most of the soil being used for GMOs is like what we inherited here.

"These companies, it's pesticides they want to sell, not food," Ross said. "Theirs is not a farm system designed to feed the world, it's a system designed to sell chemicals."

Ross takes his science very seriously. A member of the Canadian Institute for Advanced Research, he applies to soil science the same research instincts he once used as a professor of geology. He pays intimate attention to soil bacteria, and erosion control, and the symbiotic relationship between nitrogen-fixing microbes and the nodules on the roots of plants like sun hemp. Once he figured out that increasing the organic matter beneath his crops by just 1 percent saved 19,000 gallons of water per acre, he started collecting and composting 25 tons of local food waste every year.

"We're sequestering CO_2 like you wouldn't believe," Ross said. "If you're an earth scientist, you understand that with systems, things work together. Plants and microbial rhizomes, that's a 400-million-year-old relationship. Why would we trash that? Why not use 400 million years of evolution instead of fifty years of pesticides?

"This is why I have such a big problem with GMOs—it's not taking the time to understand natural systems," Ross said. "There's no freaking way a Bt gene should be in corn. There's a certain element of human arrogance. We used to be told 'one gene, one trait.' Now we have epigenetics telling us that echoes can be felt four generations down the line."

Like Alika, Ross does a lot more than farm. For years, he taught sustainable agriculture at a nearby learning center for children whose lives, one way or another, had gotten off track. Six times a year, he brought them to his farm to learn science and farming: he taught them about the structure of seeds, how seeds make plants, how plants make food. Mainly, though, he taught them "to learn that they're not stupid."

To generations of young people, Ross became known as Farmer Gerry. Years later, when the GMO debate started to get hot, these allegiances would prove critical. Young people would come out to vote, many of them for the first time in their lives.

Who Cares for the Land—the Companies or the People?

For indigenous farmers like Alika Atay and organic farmers like Gerry Ross, the GMO issue brought old legal debates over land sovereignty to the surface. Hawaii is one of the few states in the country with environmental stewardship written right into the state constitution's "public trust" doctrine. When Hawaii held a state constitutional convention in 1978–1979, the land stewardship language remained.

> For the benefit of present and future generations, the State and
> its political subdivisions shall conserve and protect Hawaii's
> natural beauty and all natural resources, including land, water,

air, minerals, energy sources, and shall promote the development and utilization of these resources in a manner consistent with their conservation and in furtherance of the self-sufficiency of the State. All public natural resources are held in trust by the State for the benefit of the people.

Alika considered Maui's anti-GMO movement to be directly connected to this tradition. "For land and water to be protected as a public trust, for animals and birds and fish to have rights, and most importantly for kids and elders to have health—if you were raised here, you're bound to those core values," he said.

Autumn Ness was not born or raised on Maui, but she knew a threat when she saw one. Ness had spent twelve years living in Japan, where, in 2011, she worked for tsunami relief efforts and set up testing facilities after the Fukushima nuclear disaster. When she became pregnant, she sought a home away from the radiation and chose Maui.

It was a safe bet, she thought, until she came across photographs of handwritten company pesticide spray logs. *July 10, 2014, 10 a.m.: Permethrin. July 15, 2014, 10 a.m.: Lorsban. July 15, 2014, 10 a.m.: Penncap. August 5, 11:30 a.m.: Malathion.*

"When I saw the spray logs, my heart sank," Ness said. "That's when I said, 'Okay, I'm all in.' Those logs came from the *least* secure fields. Other fields are triple barb-wired, like you're crossing the border between Israel and Palestine. What they were doing on fields you can walk to makes you wonder what they're doing behind all that barbed wire."

Ness turned to every authority she could think of to find out more about the chemicals being sprayed on the experimental plots and—as people had on Kauai—always came up empty. "Overshadowing all the issues is the fact that the corporations have hijacked every level of our government," she said. "That's a way bigger issue to me than the spray thing.

"The cards are institutionally stacked against us, and it's done in a really dishonest way," Ness said. "Everywhere we turned—looking for spray records, or birth defect records, or records of companies spraying near schools—we would get told by every level of people— the workers, the city council, the Department of Health, the Department of Ag—they would all say, 'I understand your problem, but there's nothing I can do for you.' I have to wonder: Who the hell is running the show here? Everyone is giving the companies the keys. Even the judges—we hear, 'I can't do anything for you.' I mean, come on! You're a judge!"

Together with Alika's SHAKA Movement, Ness became a central figure in a campaign to get a measure on the county ballot that would put a moratorium on all GM farming until the companies performed full health and environmental safety tests. The original draft included page upon page enumerating the reasons GMOs and their associated pesticides were unwelcome on Maui. The experimental plots were not farms but "an outdoor laboratory" that promoted intensive pesticide spraying on Maui and encouraged "527 million pounds of additional herbicides on the nation's farmland." The overuse of pesticides damaged soil, wildlife, and drinking water, all of which have "cultural and spiritual significance" to the island's indigenous community. GM crops constituted an "invasive species" that threatened the island's delicate balance of native plants and animals, and the pesticides used to grow them posed health risks to both consumers and farmworkers.

The petition also urged voters to consider the "Precautionary Principle" that the U.S. Supreme Court articulated in 1986: federal law mandated that states could not "sit idly by and wait until potentially irreversible environmental damage has occurred or until the scientific community agrees on what [environmental risks] are or are not dangerous before it acts to avoid such consequences."

Autumn Ness got busy knocking on doors. As part of her signature-gathering campaign, she carried along the pesticide spray logs, both as she talked to voters and when she was interviewed in the press. She published them on Facebook and in the newspaper. She circulated satellite images of the island, with experimental fields outlined in red and dramatic yellow and blue arrows indicating the direction in which chemicals would drift into towns through the air or in creeks and rivers. The images, intentionally or not, resemble military target maps, with the arrows passing directly over elementary schools and wildlife refuges.

"These guys were spraying many times a day," she said. "It's not farming, it's chemical testing. As soon as the companies found out those photos were a central part of our campaign, they went back and ripped down the board where they had posted the spray records."

Ness needed to work quickly. In order to place the measure before the county council (which could either vote on the referendum directly or pass the measure on to voters instead), organizers needed to gather 8,000 signatures within six months.

By the end of May, with Alika organizing farmers and people in the indigenous community, and Ness knocking on doors, they had collected more than 11,000 signatures in just six weeks.

The ball was now in the county council's court. During a series of "excruciating" hearings on the measure, people from both sides of the debate showed up to pressure the council. Monsanto organized a rally in front of the Maui County Building. Workers showed up wearing neon yellow hats and T-shirts and carrying signs emblazoned with "Save Ag Jobs" and "Save Farmers." "I think the initiative will threaten not only agriculture, but a lot of great jobs for the people of Maui," a worker named Lowella Oasay told a local reporter.

A Monsanto employee named Carol Reimann appeared on a video delivering "over a thousand pages of weighted studies and documents and research papers that attest to the health and safety of our products and farming practices in Maui County." A man wearing a neon yellow shirt with a Monsanto emblem on the breast said, "I love the research, I love what I do, I love working in agriculture. I've been doing it for seventeen years. It's still what drives me. I know what we do here has an impact around the word, you know, and that's important to me. That's why I do it."

Another Monsanto employee, Dan Clegg, said the documents were evidence of the company's "transparency." "I don't want to speculate, but I would say there is a group of people that have signed that petition that are thoroughly confused," Clegg said. "They don't have all the information. Now is their opportunity to step back, think about where they want local agriculture to go, get all the information before they make a decision. This is one-stop shopping, okay, for a global round of information."

Autumn Ness was impressed—and embittered—by the company's tactics. "Workers were bussed in from Monsanto and Dow— and they all said, 'If this passes, I'll lose my job,'" Ness said. "They all had their testimony written on Monsanto letterhead. For many of them, English was their second language, yet they all used the same colloquialisms. It was obvious that the same person had written their speeches."

In the end, the council declined to vote on the bill outright. The GMO ban became the first voter initiative in Maui's history to make it onto a ballot.

For the industries confronting the ban, things suddenly got serious. Stopping the GMO ban was no longer a matter of twisting a few arms on the county council; now the companies' global business model would be up to the whims of the people of Maui themselves. The companies "didn't have any idea we would get as far as we did," Ness said. "They ignored us for a while. There wasn't a peep. Then right about when it became clear we were going to get on the ballot—it's really hard in the state of Hawaii to do that—the companies were like, 'Oh, shit! Now we have something to deal with!'"

The Counter Campaign

The companies reacted swiftly. Monsanto and Dow AgroSciences were determined not to let Maui become another Kauai. Monsanto vowed to mount "an aggressive campaign against this initiative," company spokesperson Dawn Bicoy said. Banning GMOs would "devastate our county's fragile agricultural economy." The initiative, Monsanto claimed, was based on "false claims that are not supported at all by the overwhelming body of scientific evidence." GM crops are "critical to making food available and affordable to the world while also protecting crops threatened by disease, like Hawaii's own papaya."

Rather than try to convince Maui voters of the safety of GMOs, the companies tried to change the debate; instead of talking about pesticides or land rights or local produce, they would talk about jobs. The bill was not a ban on "GMOs," it became a ban on "farming." Monsanto and Dow said they employed more than six hundred workers on the island and said the GMO ban would put local farmers out of work. But they also returned to the old playbook: GMOs were necessary to feed the world. "With almost 18 million farmers worldwide growing genetically engineered crops—90% of whom are small farmers in developing countries—the SHAKA Initiative would stop Maui farmers from taking advantage of modern technology to help address some of the most pressing problems facing agriculture today," the Hawaii Crop Improvement Association, an industry group, said.

A letter, composed on letterhead from the Citizens Against the Maui County Farming Ban, went out to all registered voters. A petition was circulated asking voters if they supported a ban on farming. Ads began appearing on television and the radio, never mentioning GMOs—or that the funding had been provided by Monsanto or Dow.

The companies also flexed their muscles on the wording of the ballot initiative itself. Ness and the rest of the ban's supporters assumed the bill's ambitious language (with its references to the "spiritual significance" of the island's water and land) would be what voters would see on the ballot. This proved to be naive. By the time the ballot measure emerged from the county clerk's office, its language was so muddled that even supporters could barely understand what they were being asked to vote for.

VOTER INITIATIVE: GENETICALLY ENGINEERED ORGANISMS

Should the proposed initiative prohibiting the cultivation or reproduction of genetically engineered organisms within the County of Maui, which may be amended or repealed as to a specific person or entity when required [for] environmental and public health impact studies, public hearings, a two thirds vote and a determination by the County Council that such operation or practice meets certain standards, and which establishes civil and criminal penalties, be adopted for Maui County?

"When I read it, even I didn't know if *I* was going to vote for it," Autumn Ness said. "There was no mention of the moratorium. They changed 'GMO' to 'GE.' They did everything they could to make people *not* understand the question on the ballot. They said if we didn't like the wording, we could sue, but then we would have had to wait until the next election. So we said we would just take it. In the end, we realized we were working against our own government. We just decided we would go out and educate people."

To Gerry Ross, the influence the big companies had on local politics became clear during a meeting of the Maui County Farm Bureau. Ross had served on the farm board for fifteen years, and relationships between small organic farmers like him and the "corporate guys" had usually gone pretty well. But one evening, about eight months before the GMO vote, the corporate guys started talking about how the anti-GMO people are all "anti-science." Even the mayor parroted this line, saying that people had been "genetically modifying food for 10,000 years."

This did not sit well with Ross.

"I said, 'Wait a minute, Mr. Mayor,'" Ross told me. "'We've been *selecting* seeds for 10,000 years. We've only been genetically *crossing* for forty years. What would you do if you learned in 1959 that a chemical like atrazine actually turned out to be much more dangerous, and at lower levels, than you first supposed? You really need to study how safe this stuff is.'

"That's the kind of stuff a small-town mayor doesn't understand."

Ross agreed to add his voice to television and radio spots supporting the GMO moratorium. He went back into character as "Farmer Gerry," hoping to reach his former students—now grown up and ready to vote—to get their friends and families to show up at the polls.

Autumn Ness helped organize some four hundred volunteers and set out again to talk to her fellow islanders. There was a lot of ground to cover, especially since it was clear the companies were about to drop a lot of money on the campaign. Going door-to-door, it became clear to Ness that "nobody knew what a GMO was," she said.

"Right out of the gate the companies turned this into a farming ban," Ness said. "We were out in the community talking to people, and they thought there were two things on the ballot: GMOs and a farming ban. People told me they were going to vote yes on the GMO ban and no on the farming ban—and there was no farming ban.

"People at the door would be a strong no, then we'd have to talk to them at the door and tell them the info they had was wrong—even if the ad was on TV. We could flip a no to a yes at the door in five or ten minutes. All they needed to know was the truth."

Ness figures she alone spoke to 3,000 people. She and her team handed out fliers, reminding voters that the moratorium was intended to stop just GM experimentation, not traditional farming. GM farms represented only 1 percent of Maui's 852 farms (and just 6 percent of the island's 54,500 acres of cropland), and almost all the locally grown food people actually ate had nothing to do with GMOs. Local produce farms—farms that produced food that local people actually ate—would not be affected.

Controlling the Airwaves

In September 2014, the companies' multimillion-dollar media wave crashed over the island. Legally, Monsanto and Dow could buy only four radio commercials per hour, so that is what they did: four per hour, every hour, per station, Ness said.

"The TV and radio commercials started, and they were just relentless," Ness said. "There was no limit to airtime on TV, so they bought up every available space on TV. So even if we did raise money for ads, there wasn't any space available. By the time we got some money together, we could only buy spots at eleven p.m."

Industry advertisements—typically attributed to the Citizens Against the Maui County Farming Ban—flashed photographs of farmers working in cornfields. Voice-overs from the head of the local farm bureau emphasized that farming helps "contribute to the economy, provide jobs, pay taxes, and maintain the land in an environmentally friendly way." The companies simply "bring in supplies" that help local farmers "reduce their cost of production."

The ads "never once mentioned the safety of GMOs, they never talked about toxic chemicals. What they did talk about was a farming ban, and what agriculture means to Maui," Ness said. "The companies got older people, who remembered the plantation days, and told them if the moratorium went into effect, their families would lose their jobs. These Dow and Monsanto reps don't go on TV. They got local people to go on TV and tell their sob stories. It was crazy—really, really intense how emotional it got. They really pulled the heartstrings. They put a Filipino girl on TV saying, in tears, 'I don't know how we're going to pay our rent and our kids doctors' bills.' It even got to me."

Weeks later, when campaign finance reports came in, the financial power of the companies became clear. The industry group Citizens Against the Maui County Farming Ban received $5.1 million from a "citizen" named Monsanto; $1.7 million from Dow AgroSciences; and $1 million from the pro-industry Council for Biotechnology Information.

The campaign finance reports themselves were absurdly opaque. The companies spent hundreds of thousands of dollars on direct mail and millions of dollars on media advertising. But they also spent thousands of dollars on "media training" and a "Maui County Farm Fair" and "committee meeting prep" and "sign waving." During a parade at the Maui County Farm Fair, Monsanto employees showed up in large numbers, waving signs; several told Alika the company had paid them $200 to march behind a tractor.

All told, companies and their lobbying arm spent more than $8 million on a county ballot measure. "For all the money they spent, they could have done the safety studies and the soil testing and the water testing, and been back in business for way less than that," Ness said.

To Alika, the tactic of using workers to push a political agenda was doubly distasteful.

"The companies always dangle the carrot of money," Alika told me. "For me, when you step away and look at it, the real issue is this: There are those who live here, and those who just sleep here. A large majority—maybe 80 percent—of the workers on these farms are immigrant Filipinos or Micronesians; they're international migrants. Yeah, they have families, but they're here on work visas. So when I ask them, 'Where are you from? Where is your home?' the Filipino guys say, 'I send all my money home'—meaning back to the Philippines. But then they get paid by the companies to show up at rallies. They had two hundred of them show up at a rally at the county fair, and the guys told me they were paid to be there. They show up at county council hearings, same way.

"The same thing is true for these big, burly tractor operators from Nebraska. They just sleep here. They come and go. They come here when it's snowing back home, go back when it's warm. Even the scientists—they come from places like France, so they just sleep here too. All these people saying that GMOs are so good—this isn't their home. For us, this is our home. I ask the Filipinos: 'If they sprayed five times a day in your county, what would you do? Why is it okay to poison us?' I don't blame the workers, I blame the economic system that has them working here in the first place."

Lorrin Pang, a Maui physician and a consultant to the World Health Organization, maintained throughout the campaign that he was "very concerned" with the experimental GM crops, especially because of the chemicals they required. "You may know the effects of each chemical individually, but each new combination could have stunning effects," he wrote. "The minute you combine then, all hell can break loose. We've only recently learned that, on Kauai for example, they are regularly spraying seventy to eighty different chemicals to kill everything in the soil, the microbes, the viruses, the fungi. That represents ten to the twenty-third possible combinations, a trillion trillion, more than all the drops of water in the ocean. And they certainly haven't cleared any of this with the people who have to live with the risk of being exposed to whatever is being tested. This is all quite unethical."

With election day approaching, celebrity anti-GMO activists started showing up to support the effort. Here was Tyrone Hayes, the Berkeley biology professor and former Syngenta scientist who made international headlines for showing that Syngenta's pesticide atrazine causes hormone disruptions. There was Ben Cohen of Ben & Jerry's Ice Cream, speaking about how difficult it was to be GMO-free in the ice cream industry "because most of the feed given to cows comes from GMO crops."

Alika and the SHAKA Movement held a daylong Hawaiian music festival called Aloha da Vote. A party called Shake It for SHAKA advertised "tribal ethno global beats to move feets & stir us into ecstatic bliss dance heaven."

The Vote

Alika Atay didn't care whether they danced, walked, or drove to the polls, he just wanted to get them there and get them to vote. The early returns did not seem promising. Local television and radio stations continued to bombard Maui residents with ads paid for by Monsanto and Dow, and the tactic seemed to be working: as Alika drove around, he noticed the polling places were empty. With just four hours left before the polls closed, exit interviews indicated that the industry side was winning 60–40.

Alika and his team began feverishly working Facebook and Twitter. They called everyone they knew. If you haven't voted yet, get out and vote. If you have voted, fill your car with friends who haven't, and get them to the polls.

"We had people working all the precincts," Alika said. "We said, 'Let's make our signs the last things people see before they vote.'"

Autumn Ness said she was never in any doubt. She knew how many doors she had knocked on. Sure enough, when the final vote was tallied, supporters of the moratorium—a shoestring, grassroots organization battling $8 million spent by two of the biggest companies in the world—had won, with just over 51 percent of the votes. The vote to ban all GM farming on the island was decided by just a thousand votes.

"That night, when people read the results and the reality sank in that we had won, there were a couple thousand people gathered, hugging each other," Alika told me. "I saw a lot of young people, a lot of Hawaiians, coming up to me and saying this was the first time they had ever voted. There were people who had given up on the system—the elders—they chose this time to say, 'Maybe this will be worth it.'"

The celebrations were short-lived. SHAKA and the rest of the moratorium's supporters knew the companies would take their victory to federal court, just as they had on Kauai and on the Big Island. So as soon as the votes were counted, they filed a lawsuit—unusual for the side that won an election—seeking to force the county to enforce the ban.

The next day, Monsanto and Dow Chemical filed their own lawsuit. Just as they did after the Kauai and Big Island votes, the companies claimed the Maui initiative had no authority to preempt state and federal laws that already regulated GMOs. "This local referendum interferes with and conflicts with long-established state and federal laws that support both the safety and lawful cultivation of GMO plants," said John Purcell, a Monsanto executive.

Barry Kurren, the federal judge who struck down both Kauai's bid to restrict GM farming and the Big Island's own GMO restriction, issued an injunction, pushing for more arguments to be heard; the county agreed to wait several months to start enforcement.

Kurren reassigned the case to Chief Judge Susan Mollway, and on June 30, 2015, Mollway ruled that the county law was indeed preempted by state and federal law, and that the county had overstepped its authority by banning GMOs. Notably absent from her ruling was any opinion about the safety of GMOs.

> No portion of this ruling says anything about whether GE
> organisms are good or bad or about whether the court thinks the
> substance of the ordinance would be beneficial to the county.

Alika Atay, the SHAKA Movement, Lorrin Pang, and a handful of others have appealed the ruling to the 9th U.S. Circuit Court of Appeals. Their goal: Get the county to enforce the will of its own citizens.

To Alika, the victory—however compromised—represented a

profound moment in the history of his indigenous people. No longer would native Hawaiians feel intimidated by colonial economic forces, no matter how well-heeled.

"For me, that was the bigger message," Alika said. "It gave these young people a taste of victory. They knew how much hard work and sacrifice came along with that victory. So now, when future challenges come up, they'll know what to do. We were *aina* warriors."

Part Three

FRUIT

Feeding the World

Dennis Gonsalves saved an industry by redesigning the genes of a single papaya plant. Nigel Taylor is doing similar work, but he's working to protect food for an entire continent.

When I visited Taylor, I discovered him deep inside a large greenhouse outside St. Louis. He was looking wistfully over a small forest of foot-tall cassava seedlings, pawing through a canopy of five-lobed leaves. One by one, Taylor pulled up plants, looking closely at the color of the roots. He was hoping to see orange, but—all too often—he saw white instead.

Taylor moves methodically, but there is an unmistakable urgency to his work. A soft-spoken man with a gray beard and ponytail and a rich Scottish accent, Taylor is a senior research scientist at St. Louis's Donald Danforth Plant Science Center, one of the world's leading (and most well-funded) nonprofit plant research institutions. Taylor is experimenting with genetically engineered cassava, an improved version of an essential crop grown by millions of small farmers in Africa. Cassava is dense with calories, it can tolerate heat and drought, and it can be grown in depleted, marginal soil. But like white rice, cassava is also an imperfect source of nutrition: it fills bellies, but does not fully nourish bodies. Inserting genes that would make cassava more nutritious—coding plants to produce and store vitamin A, vitamin E, or iron—might solve significant health and nutritional problems for the 250 million people who depend on the crop.

Taylor yanks up another cassava. The root of this one is the color of a Creamsicle, and Taylor smiles faintly. The gold-orange hue of the root means the plant is generating beta-carotene, the same compound that gives carrots and sweet potatoes their color. Beta-carotene is essential to the body's generation of vitamin A, a crucial nutritional staple whose absence causes blindness and death in hundreds of thousands of children in the developing world. Vitamin A is found in animal products like eggs, liver, and dairy products, but in countries that don't eat much of these things—especially parts of Africa, Asia, and Latin America—reliable sources of vitamin A can be hard to come by. With the right genetic tinkering, Taylor's "golden cassava" could help solve vitamin A deficiency for the many cultures that experience it.

But first he has to get all of the components of the genome just right, and it's not just nutrition he has to address to make the crop more productive.

There are also the flies.

In recent years, cassava crops have been attacked by growing swarms of whiteflies, which serve as vectors for a pair of viral diseases called mosaic and brown streak. These pathogen-carrying insects have long been a plague, but warming temperatures, possibly caused by climate change, have helped their numbers explode. Traditionally, the only answer has been to spray plants with pesticides, an only marginally effective solution that carries its own dangers for both farmers and the people they feed.

"Spraying to control whiteflies is not effective, because—like spraying for mosquitoes to get rid of malaria—you have to kill every one," Taylor said. "These flies are incredibly efficient; you can find a couple thousand flies on a single plant. When we were doing our first field trials, they were flying up and we were breathing them in, wheezing them in. It was really unpleasant."

In the 1990s, scientists working across sub-Saharan Africa focused on breeding cassava to develop plants resistant to the mosaic virus. They were very successful, Taylor said.

But then the brown streak disease came along.

Brown streak had been around in coastal Kenya and Mozambique for a long time, but it started spreading like crazy in the early to middle 2000s. Cassava varieties that had been developed to resist the mosaic virus were helpless before brown streak, which morphed from being an isolated disease to an epidemic throughout coastal East Africa.

"People have been looking for sources of resistance to the brown streak disease, but so far, it has proved difficult," Taylor said. "When a plant gets infected, it can recognize the pathogen, and this stimulates its defense mechanism. But when it's a battle between the plant and the pathogen, brown streak always triumphs."

Cassava is "vegetatively propagated," meaning farmers take stem cuttings from one season's crop to establish the next. Therefore, if one year's plants are infected with the disease, it is carried over to the next planting cycle. "Even with no new infections, your yields are being affected," Taylor said. "Diseases are always there. Insect vectors are always there."

Where traditional breeding is facing challenges, Taylor is counting on genetic engineering to succeed. Like Dennis Gonsalves, a scientist Taylor very much admires, Taylor is hoping to take an existing cultivar and introduce new gene sequences that—if he can get the sequences right—will make cassava resistant to brown streak. In this, his work is very much like that done on Hawaiian papaya. The difference is that on Hawaii, the price of failure is the collapse of a local industry. In Africa, the collapse of cassava would dramatically affect the lives of millions of people.

Paul Anderson, one of Taylor's senior colleagues, has something of a cold-eyed view of the interaction between humans, food, and agricultural technology. Anderson is the director of the Danforth Center's Institute for International Crop Improvement and oversees the center's work on cassava, sweet potato, sorghum, and cowpeas. He has long studied the rise and fall of crops and human civilizations, and when it comes to the human dependence on farming, he has little patience for sentimentality.

"Human populations rise and fall based on the promise of food produced in those geographies," he said. "There are lots of instances of crops going by the wayside due to various problems, and others arising. This is why some societies succeeded and some did not. One of the key factors was the ability to grow crops, and those that created multiple crops succeeded. Those that didn't were doomed to be hunters and gatherers.

"Historically, starvation typically arises with too much dependence on one type of crop," he said. "The potato blight in Ireland—that sort of scenario has played itself out in a lot of different places and in different times. Sometimes diseases could be addressed with cultural practices, with farmers noting that some things you did decreased the possibility of disease. You could manage to get by. But that sort of thing takes time. You gotta be really lucky, or get somebody already doing that cultural practice. One always tries to grow a crop where it hasn't been grown before, to find how it is limited by temperature or water availability or what have you, so any plant breeder is going to be working on expanding the value of that acre by growing in many places and having it yield well."

Take sorghum and corn. Sorghum tolerates drought quite well in places like the Sahel, the semi-arid band of Africa south of the Sahara desert. Corn (also known as maize) does not. But maize has advantages that sorghum does not: it tastes better, and its nutrients are more readily available. With maize porridge, your body absorbs 80 to 90 percent of the grain's protein, Anderson said. With sorghum, it's only 65 percent.

"Over the last ten years, more and more people are growing maize, but it is not a stress-tolerant crop," Anderson said. "But farmers really like it, so if they get good growing conditions for two, three, four years in a row, they increase the maize on their farm. But then there will be a drought, and the maize crop will fail.

"So that happens, and farmers are used to that," Anderson said. "But if it happens two years in a row, the farmers are lost. He leaves the farm and moves into the city. This has happened most recently in Kenya, after a significant drought caused big population movements. The choice of the wrong crop caused a lot of farmers to fail."

So genetic engineers have a couple of options, Anderson said. They can work on drought-tolerant maize, which plant breeders have been pushing for as long as recorded history, or they can develop a sorghum that is more palatable and has improved nutrition, Anderson said.

"Genetic engineering isn't an end in itself, it's just a crop-improvement practice that extends your ability to make improvements," Anderson said. "So depending on what time in history one was in, one had tools one could use. Genetic engineering very recently added a new tool—a significant tool, but it's no different than other tools, like the fertilization of plants, or the hybridization of corn."

The United Nations estimates that the world will be inhabited by another 2 billion people by 2050, half of them born in sub-Saharan Africa, and 30 percent in South and Southeast Asia. All of these places are projected to experience acute and worsening drought, which may well make the breakdown of food systems one of the most dangerous effects of climate change.

With such catastrophic changes on the horizon, the need for advanced technology like GMOs has never been so acute, Anderson said. "Making plants more stress-tolerant—these are difficult issues to address," he said. "It boils down to this: Is there sufficient genetic variation in the crops of interest? If not, then one has to create variation in the crop so it can be manipulated. Drought tolerance, cold tolerance—these have been targets for plant breeders for thousands of years. Genetic engineering is going to be required to make these big changes."

To Anderson, using GM technology to improve crops in the developing world is a solution that ripples far beyond the growing of food.

"In most limiting situations, you're talking about the ability to provide nutrients and calories to get you through the year," Anderson said. "You don't have to go very far to see that if you double this, or even increase it by 50 percent, you can sell your crops or share them. You can get the leverage that allows you to move out of poverty. It's poverty that's the biggest problem in these situations.

"Food availability is more dramatic, but it's ongoing poverty that won't allow a person to achieve their potential. Field labor is almost entirely women and children. Fix this, and a farmer's kids might get to go to school or have a book when they go to school."

Paying for Orphans

With so much at stake, and with genetic engineering offering so much promise, why haven't multinational corporations put more muscle into this work?

The answer is money. Or, rather, profit.

The Danforth Center looks like a hybrid between a university and a corporation, and in a way it is: the center's campus is massive, gleaming, and growing, with 200,000 square feet of gorgeous, state-of-the-art laboratory buildings set off by a sky-lit atrium and a lengthy, fountained reflecting pool. This will soon be joined by $45 million of additional research space and another hundred additional researchers—including the University of Delaware's Blake Meyers.

The Danforth Center's work is also situated somewhere between university research and corporate agriculture: they do basic science, but they also get their plants out into the field. Their work is not just theoretical, in other words; it is meant to help make practical changes in some of the neediest parts of the world. Most academic scientists are more concerned with publishing research papers than implementing full-scale field tests, Nigel Taylor said, and in any case don't have the money or the staff to deal with things like international bureaucracy, which can kill imaginative projects before they ever get off the ground.

On the other hand, global food companies, with their deep pockets and their eyes on huge profits, have almost exclusively focused their attention on commodity crops—corn, soy, canola—that make them billions of dollars a year in the enormous North American food market. Building laboratories for genetic engineering is expensive, the companies say, and they need a return on their investment to make the whole thing worthwhile. "Orphan crops"—so named because of their neglect by big industry—are left to university researchers and nonprofit centers like Danforth. Cassava, papaya, millet—these crops may be critical staples for millions of the world's poor, but they will never generate the kind of profits demanded by multinational corporations.

Instead, companies donate money to nonprofit researchers doing this sort of work: the Danforth Center's cassava project alone has received some $20 million in grants from Monsanto, as well as from the Gates Foundation and the U.S. Agency for International Development. The nonprofits get research money, and—in the bargain—the multinationals can say they are doing their part for the needy.

In other words, the Danforth Center sits at the very joint of the GMO debate: its scientists are working to help the world's most vulnerable people, but they also provide excellent public relations for companies like Monsanto to boast that GMOs are "feeding the world" and not just "feeding the fast-food industry." The relationship between the two institutions is distinct and blurry at the same time. The Danforth Center was built literally across the street from Monsanto's world headquarters in St. Louis, and both Monsanto's president and its former chief scientist sit on the Danforth Center's board of directors. Scientists move back and forth between industry and the center. Paul Anderson, for example, spent ten years as the research director of food and feed research at Pioneer Hi-Bred, the same DuPont company caught in the fierce GMO debate on Kauai and Maui. Before that, he served as a senior manager in Pioneer's efforts to move the company's grain into Asia, Eastern Europe, and South America.

A cynic might claim that the $20 million Monsanto throws to the Danforth Center is barely a (tax-deductible) rounding error compared with the company's nearly $16 billion in annual sales. More cynical would be the view that developing drought-resistant

GM corn for Africa is really just a way for seed companies to gain more influence—and market share—on other continents. The principal beneficiary of America's foreign assistance programs has always been American companies, the U.S. Agency for International Development has said. Close to 80 percent of the agency's contracts and grants go directly to American firms. "Foreign assistance programs have helped create major markets for agricultural goods, created new markets for American industrial exports and meant hundreds of thousands of jobs for Americans."

The Gates Foundation, which spent close to $500 million on African agricultural development from 2009 to 2011 alone (and which also supports the Danforth Center), has become "a stalking horse for corporate proponents promoting industrial agriculture paradigms, which view African hunger simply as a business opportunity," writes Phil Bereano, a professor emeritus of public policy at the University of Washington. Bereano calls this "agroindustrial philanthrocapitalism"; GM crops, he says, "threaten conventional and organic production as well as the autonomy of African producers and nations."

Marion Nestle, a prominent food scientist at New York University, has long been suspicious of industry's humanitarian claims. If giant seed and chemical companies really want to help "feed the world," they should dedicate substantially more resources to helping local farmers in Asia, Africa, and South America develop crops that might only be of *local* value—even if they don't promote industrial agriculture, and even if they have no potential for corporate profit. How much should companies dedicate to humanitarian food development? Nestle's modest proposal: 10 percent of annual corporate income, a kind of tithing to help those in need.

"If companies are going to claim that their work will solve world food problems, they need to put substantial resources into working with scientists in developing countries to help farmers produce more food under local conditions," Nestle writes in her book *Safe Food: Bacteria, Biotechnology, and Bioterrorism.* "I continue to believe that to be perceived as credible, the industry must *be* credible."

Indeed, given the destruction that industrial agriculture has done to the American landscape, why should we expect anything different once its technologies are exported to the developing world? "In the United States, we've seen the number of farms drop by two-thirds and average farm size more than double since World War II," wrote veteran food activists Peter Rosset, Frances Moore Lappé, and Joseph Collins. "The gutting of rural communities, the creation of inner-city slums, and the exacerbation of unemployment all followed in the wake of this vast migration from the land. Think what the equivalent rural exodus means in the Third World, where the number of jobless people is already double or triple our own."

This kind of criticism drives scientists like Paul Anderson crazy. Critics of GMOs, especially those in food-secure places like the United States and Europe, have no idea what's at stake for the lives of the poor, he says. Anderson takes an especially dim view of what he calls "anti-technology groups that are funded by Europeans."

"Do they have the right to do that? Do they have the right to decide who is going to eat what?" Anderson said.

Nigel Taylor agrees. His work on virus-resistant cassava is unlikely ever to serve any corporate interest, and like Dennis Gonsalves, Taylor's primary interest is in serving small African farmers.

"It's important that African farmers have a strong say in this because it's their livelihood, and they should have the right to access any technology that can improve their standard of living," Taylor said. Creating a virus-resistant cassava plant "would be highly desirable because of the seriousness of brown streak to people's economic security in East Africa."

Given all the noise involved in the GMO debate, Taylor would plainly prefer to leave politics aside and simply work with his cassava plants. He leads me into his tissue culture laboratory to show me minute cassava embryos—clusters of totipotent cells that will be genetically altered before being grown into fully developed plants. Once altered, the cassava cells, under the watchful eyes of Taylor's team, can be cultured and turned into a thousand or more plants. Of these, only a small number will carry the genetic material needed to protect the cassava plant from brown streak. Much work is required to identify and select the most efficacious. The introduced virus defense works by enabling a plant to recognize a viral infection before it occurs, which it does by generating small RNAs and proteins known as argonauts (named for the Greek explorers) that act to "silence" the infecting virus.

"What we can do by triggering this defense mechanism early—it's not an immunization, but it is similar in the manner that it's pre-arming the plant's defense mechanism," Taylor said. "As the virus replicates, it makes double-stranded RNA. The plant can recognize that, and its inherent RNA-silencing mechanisms grab it and chop it up, preventing establishment of the disease. However, in the non-modified plant the virus wins the battle, as the plant cannot fire up these defense mechanisms fast enough to stop the virus replicating and moving to establish infection. By modifying the plant to recognize the virus, and activating the RNA defense mechanisms before the virus arrives, the plant will always stay ahead of the virus and will be resistant. And since the plant makes its own RNA continuously, the plant will always be resistant. So we're not making a new defense mechanism, we're just turning on the plant's inherent resistance systems early and keeping them on."

Let's say his lab creates 600 cassava plants. Two-thirds of them would likely need to be thrown out for not expressing virus resistance. Once plants have been selected in St. Louis, they are field tested in Puerto Rico, which, like Hawaii, is a popular growing environment for experimental crops. Then maybe twenty plants get to the field in Africa. These get whittled down to one or two that go through all the stages of testing within the regulatory system. Only one plant line would be submitted for formal approval. If this makes it all the way through regulatory testing and approval, countless crops of this improved cassava line could eventually be grown from this one parent plant.

In Kenya and Uganda, Taylor and his team work with African scientists and government officials. They have conducted socioeconomic studies to assess if farmers would be receptive to what they are offering.

"If you frame it up for small farmers for on-farm consumption and local trading, almost everyone says yes," Taylor said. "This is such an important disease, a major threat, and there are very few ways of addressing it. If we can show this works, the farmers have indicated that they would be ready to adopt it."

The question for the people at the Danforth Center is whether their cassava will turn out to be a hit, like Dennis Gonsalves's papaya, or a misfire, like golden rice.

Golden Rice: The Grain That Will Save Millions of Children—or Won't

In the summer of 2000, *Time* trumpeted a cover story about a GM grain that it said could "save a million kids a year." The magazine featured a cover photo of Dr. Ingo Potrykus, a gene scientist who had spent ten years trying to alleviate the suffering of millions of children in the developing world who have deficient levels of vitamin A. Lack of this single nutrient causes blindness in up to half a million children each year and weakens the immune system to the point that some 2 million people die each year of diseases they would otherwise survive.

How to get more vitamin A into the Asian diet? Potrykus had developed what seemed like a brilliant solution: by inserting genes from daffodils into a rice genome, he had derived a plant fortified with beta-carotene, the same pigment Nigel Taylor is hoping to introduce into cassava. Asia alone produces 417 million tons of rice a year; the trouble is, even if children in many poor countries can get their hands on rice, they frequently do not have access to vitamin-rich fruits or vegetables.

Potrykus visualized peasant farmers "wading into paddies to set out the tender seedlings and winnowing the grain at harvest time in handwoven baskets," *Time* reported. He pictured "small children consuming the golden gruel their mothers would make, knowing that it would sharpen their eyesight and strengthen their resistance to infectious diseases. And he saw his rice as the first modest start of a new agricultural revolution, in which ancient food crops would acquire all manner of useful properties: bananas that wouldn't rot on the way to market; corn that could supply its own fertilizer; wheat that could thrive in drought-ridden soil."

Even more than Dennis Gonsalves and his GM papaya, golden rice made Potrykus and his research team international celebrities, not least because, like Gonsalves, they had done their work for a nonprofit institution—the International Rice Research Institute (IRRI), based in the Philippines. The project got $100,000 in seed money from the Rockefeller Foundation, and another $2.5 million from the Swiss government and the European Union. But because nearly six dozen genes they were interested in had already been patented by some thirty-two companies, the research team also had to tiptoe through a legal swamp. DuPont, Monsanto, and Zeneca owned a piece of the rice genome, as did Stanford and Columbia universities and the universities of Maryland and California. Patents to the daffodil genes were held by Amoco, DuPont, Zeneca, and Imperial Chemical Industries. The patent for the bacterium was held by Japan's Kirin Brewery.

In the end, Potrykus and his team struck a deal with AstraZeneca (now Syngenta) that gave the researchers the rights to the seeds they would give to farmers in developing countries earning less than $10,000 a year. The company retained the right to market the rice in places like Japan and the United States. To its supporters, this seemed an ideal partnership between public scientists and private industry, especially after other corporations holding patents also waived their own rights.

The journal *Science* announced the successful experiment by distributing magazines to 1,700 journalists around the world. In an accompanying note, editors claimed that "this application of plant genetic engineering to ameliorate human misery without regard to short-term profit will restore this technology to political acceptability."

Indeed, whatever golden rice's prospects for the world's poor, the announcement was a spectacular gift for the biotech industry. After being battered by nearly two decades of growing public resistance to GMOs, biotech companies jumped at the chance to boast that genetic engineering would now feed the world.

The backlash came swiftly.

"A rip-off of the public trust," grumbled the Rural Advancement Foundation International, an advocacy group based in Winnipeg, Canada. "Asian farmers get (unproved) genetically modified rice, and AstraZeneca gets the 'gold.'"

Greenpeace, which had taken a strong stand against GMOs

from the beginning, mocked golden rice as an intentional ploy to reverse public anxiety about the technology. "People are talking about the potential benefits of the second generation of genetically modified crops when almost no questions raised by the first have been answered," the group announced. "You don't have to be paranoid to think the tactics are deliberate."

In an article in *The New York Times Magazine* titled "The Great Yellow Hype," journalist Michael Pollan suggested that golden rice was being exploited by the biotech industry "to win an argument rather than solve a public-health problem." Malnourished children would have to eat fifteen pounds of cooked rice a day to satisfy their nutritional needs, Pollan wrote, and even if they could eat that much, their fat- and protein-deficient diets would prevent their bodies from taking up the beta-carotene.

"The unspoken challenge here is that if we don't get over our queasiness about eating genetically modified food, kids in the third world will go blind," Pollan wrote. "Granted, it would be immoral for finicky Americans to thwart a technology that could rescue malnourished children. But wouldn't it also be immoral for an industry to use those children's suffering in order to rescue itself? The first case is hypothetical at best. The second is right there on our television screens, for everyone to see."

And Vandana Shiva, who would become an international celebrity for vehemently opposing golden rice, called the grain a Trojan horse for the biotech industry. In books like *The Violence of the Green Revolution*, Shiva had lambasted the planting of Western varieties of wheat and the attendant herbicides, which pushed traditional, vitamin-rich greens like bathua to extinction. Now, she wrote, "the 'selling' of vitamin A as a miracle cure for blindness is based on the (corporate) blindness to the alternatives."

And so it has gone. Even in countries where vitamin A deficiency has been most acute—where, one would think, support for such a product would be uniformly enthusiastic—golden rice has been met with acute skepticism and even violence. The government of India is still considering banning all GM field trials for ten years. In Kenya, the government has banned the import of GM food (though not GMO research).

In August 2013, hundreds of protesters smashed through fences surrounding a field in the Philippines so they could uproot a plant that had been hailed as the potential savior of millions of Asia's malnourished poor. "We do not want our people, especially our children, to be used in these experiments," a farmer and leader of the protest told the Philippine newspaper *Remate*.

To this day, golden rice—once seen as a savior of the global poor—has not been approved by a single country. What happened?

The Green Revolution

The International Rice Research Institute, where Potrykus did his work, had been launching successful breeding projects for decades, and until it started working with GMOs, it had largely met with global gratitude. In the early 1960s, a plant pathologist named Peter Jennings created a fast-growing, high-yielding strain known as India Rice 8 that became so popular that (legend has it) some Indian families even named their children "IR8."

Such research mirrored work generated by other scientists at the center of what came to be known as the Green Revolution, which used both new plant-breeding techniques and the heavy use of petrochemical fertilizers, pesticides, and herbicides to dramatically increase the amount of food that farmers could grow. Between 1950 and 1983, crop yields of cereal grains doubled, tripled, even quadrupled. Since grains provide about 80 percent of the calories people consume worldwide, such advances dramatically improved the diets of billions of people: between the 1970s and 1980s, the total amount of food available per person in the world increased 11 percent, while the number of hungry people fell 16 percent (from 942 million to 786 million).

When Norman Borlaug, a researcher at Texas A&M University, won the 1970 Nobel Peace Prize for developing high-yielding wheat and rice, his citation said that "more than any person of this age, he helped provide bread for a hungry world."

Borlaug has never been shy regarding his feelings about how best to feed the poor and hungry. The organic movement is "ridiculous," Borlaug has said. "For those who want to go the organic route, God bless them. Let them spend more money for their food. But looking at the world at large, this is an impossibility. . . . Most of the people who are opposing biotechnology, they've never known hunger. These people say that the little farmer should permanently accept that he's going to stay on that three-acre farm with a hoe and a machete. That's fine in Utopia, but don't give the world the false idea that they can produce the food that's needed for 6 billion people."

Borlaug's thoughts notwithstanding, the Green Revolution, which laid the global foundation for the spread of GM crops, also left a swath of troubling consequences. The synthetic fertilizers that spurred such high crop yields also created more weeds and insects, which led to a huge increase in the use of herbicides and insecticides. In India, chemically treated land jumped from 15 million acres in 1960 to more than 200 million by the 1980s, contributing to a dramatic global increase in human exposure to toxic chemicals. These poisons also killed natural predators, and soils worldwide edged closer to becoming chemically saturated and lifeless.

M. S. Swaminathan, a renowned Indian geneticist and a leader of India's Green Revolution, later recalled that he had foreseen these complications as early as 1968. "Exploitive agriculture offers great possibilities if carried out in a scientific way, but poses great dangers if carried out with only an immediate profit motive," he said. "Without first building up a proper scientific and training base to sustain it, [it] may only lead us, in the long run, into an era of agricultural disaster."

By 1999, Swaminathan noted that "the significance of my 1968 analysis has been widely realized."

Plant geneticists like the Danforth Center's Paul Anderson believe GMOs may provide an answer to many of these global food problems. But the debate over spreading GMOs across the developing world has additional complexities, notably that the technology, and the industries pushing it, are largely based in Europe and the United States. The shadow of colonialism has not been lost on local political leaders or on anti-GMO scientists.

Indeed, for every plant scientist who sees GMOs as a powerful tool to feed the world, there is a scientist, or an activist, worried that genetic technology will simply speed up the processes of industrial agriculture that are already in place. Despite the boasts of chemical and biotech companies, there is little evidence that GM crops reduce global chemical use; rather, pushing GMOs at home and in the developing world "has contributed to the increased use of herbicides to control weeds and the resulting increase in environmental pollution," Cornell's David Pimentel writes.

Activists, for their part, have gone to great lengths to make GM crops a symbol of colonial exploitation. In 1999, the Earth Liberation Front torched a genetics research building at Michigan State University where researchers were working on crops for the developing world. The fire caused $900,000 in damage. A spokesman for the group said the research was "not going to end world hunger, it's going to make more profits for Monsanto."

Catherine Ives, the scientist in charge of the lab, was heartbroken. "I would wonder how much time has been spent by people in this organization in developing countries," she told PBS. "I see women hiking for miles to bring firewood in because they've cut down everything around them and have no productive soils. I see children who are malnourished. They do not have sustainable agricultural practices in place in many parts of the world. That is what we are trying to help them develop."

Feeding the Poor, or Expanding Markets?

Critics say the future of GMOs will play out in the developing world, and not necessarily with the benefit of local people in mind. With upward of 90 percent of American corn and soy crops already planted with GM seeds, the only place for industry to expand is in places like Africa and South America. In this view, GMOs will continue to cause social disruptions that are at least as harmful as their ecological disruptions. Just as they have in the United States, the GM soybeans spreading across countries like Argentina and Paraguay are already replacing diverse, traditional crops with less nutritious monocultures mostly being used to feed livestock for the expanding global market for beef. Just as in the United States, these larger and larger farms move from local control to industrial control.

Monsanto boasts that it has already trained 4,000 farmers in South America to use the company's seeds and pesticides. Paraguay, where 80 percent of the land is controlled by 2 percent of the population, has become the world's fourth-largest exporter of soy, with more than 3 million hectares of fields producing more than 8 million tons of soybeans a year. Most of this is GM, and all of it is heavily doused with chemicals, which has contaminated local water supplies and caused public health scares. In 2013 alone, 914 square miles of pristine forest in a wilderness known as Gran Chaco was cut down and burned to create soybean fields.

Paraguay's former president Fernando Lugo took a firm stand against global food companies, at one point ordering his own agriculture department to destroy GM cornfields. The destruction of forests and traditional farming, he argued, was ecologically disastrous and destructive to both traditional farmers and the country's indigenous people.

Two years later, in 2012, Lugo was ousted in a coup he claimed was orchestrated in part by multinational food companies. His successor, Federico Franco, fast-tracked approval of seven GM soy, cotton, and corn strains; Monsanto, whose Roundup Ready soybeans are used in 95 percent of Paraguay's production, was authorized to sell its new GM seeds in Paraguay just seven months after Franco was sworn in.

In Argentina, meanwhile, a woman whose infant daughter died because of pesticide poisoning was given the 2012 Goldman Environmental Prize, one of the world's most prestigious, for her efforts to ban agricultural spraying. Argentina is the world's third largest exporter of soybeans; industry uses airplanes to spread more than 50 million gallons of pesticides, especially glyphosate and endosulfan, over GM crops. Going door-to-door to collect stories of families whose homes were surrounded by soybean fields, Sofia Gatica found cancer rates forty-one times the national average. Despite enduring threats from police and local business owners, she persuaded Argentina's health minister to investigate. In 2010, the country's supreme court banned agricultural spraying near populated areas, and—reversing a tradition of forcing residents to prove harm—required that companies prove their products were safe. Argentina banned endosulfan in 2013. Gatica and her colleagues are now pushing for a nationwide ban on glyphosate.

It is in this pot—a world reconsidering both the Green Revolution and what GMOs might contribute to relieving (or worsening) global food problems—that the debate over golden rice remains simmering.

There are two primary strains of rice grown in Asia: the short-grained *japonica* (think sushi) and long-grained *indica* (think jasmine or basmati). In Asia, farmers have been growing countless varieties of rice for thousands of years. Each strain is developed to accommodate both natural forces (like drought and flooding) and personal taste. It's not just that a grain that grows well (and is considered desirable) in Bangladesh is different from a grain in Japan or China; there are grains preferred in individual Bangladeshi villages that are different from grains grown just a few miles away.

Here's the problem: Most of the original research done on genetically modified rice was done on *japonica*. Most of the poor in Asia eat *indica*. You can have short, fat grains that are stacked with all the nutrients on earth, but people won't eat it if they prefer grains that are long and skinny. Equally challenging, most people in Asia prefer white rice, despite the fact that it is considerably less nutritious than brown rice—let alone rice of an unrecognizable golden color.

"Rice is white, not yellow or golden, and people are very specific about what they will eat," Alfred Sommer told me. "So let's assume it's a yellow version of their local version. Assume it grows well. Will anyone eat yellow rice? Taste, flavor, what it looks like—we don't know that. We haven't tested it out yet, so we don't know. If you gave golden rice to 10,000 people, would they eat it? We have no idea."

Sommer knows a thing or two about vitamin A deficiency. So successful were his early efforts to combat the problem that his name adorns an entire wing of the Bloomberg School of Public Health at Johns Hopkins, where Sommer is a professor of epidemiology and former dean. Long before the invention of golden rice, Sommer figured out that solving vitamin A deficiency was as simple (and cheap) as getting people to swallow a couple of two-cent vitamin A capsules a year. The trick was getting the pills to the people. Early ideas included combining vitamin A with MSG or salt, but neither really worked. Sugar worked well in places like El Salvador and Guatemala, where people grow their own sugar and process it in factories where vitamin A can be distributed, but these conditions rarely exist in other countries.

Some clinics combined vitamin A distribution with polio vaccinations, but not everyone who lived in the countryside made it into a clinic. Health workers (or volunteers) would have to trudge long distances to reach remote villages.

"The problem is that there are very few products that poor people consume widely that we can get to them through some kind of central processing," Sommer told me. Inserting vitamin A into a universal consumer food like rice seemed like the magic bullet— provided it could be grown locally, would appeal to local tastes, and, critically, provide the nutrient in sufficient quantities. The original version produced a woefully inadequate 1.6 micrograms of beta-carotene per gram of rice; an improved version called golden rice 2 (and developed at Syngenta) replaced the daffodil gene with a gene from corn, and now delivers up to 37 micrograms per gram.

When golden rice was first being touted in the mid-1980s, the president of the Rockefeller Foundation, which had partially financed the research, asked Sommer to write a piece about it.

Sommer wrote—and maintains—that although considerable research remains to be done, when all is said and done, golden rice can be an extremely useful tool for combatting vitamin A deficiency.

Twenty years ago, the anti-GMO hysteria was already "irrational, but well deserved, given what Monsanto's poor PR helped to create," Sommer told me. Golden rice was an attempt to bring something to fruition that "clearly would address a major public health issue and would also help overcome the negative publicity that had been generated by Monsanto's approach to pushing GMO foods, which were not seen as beneficial to anyone but the agriculture industry."

If anything, the prospect of climate change has sped rice research generally into overdrive. In 2004, an international consortium mapped the entire rice genome; two years later, Pamela Ronald of UC-Davis isolated a gene called Sub1 that helps a plant survive even when submerged in floodwaters for two weeks (most plants die after being immersed for three days). The strain has proven popular in flood-prone places like India and Bangladesh, where some 4 million farmers now plant a version of this rice.

In 2015 alone, scientists published the genomes of 3,000 strains of rice, many focusing on drought- and salt-resistant strains that might survive hotter temperatures and rising sea levels. Researchers are also hoping to produce a photosynthetically enhanced rice grain that would increase yields 30 to 50 percent with the same amount of water and fertilizer. The Gates Foundation has given $20 million to the project, which has twenty-two teams of researchers from nine countries working on it.

On a hotter planet, with rising tides, growing populations, and diminishing supplies of fresh water, biotechnology will be more important than ever to produce enough food, especially in the developing world, the Danforth Center's Jim Carrington told me. Regions like Africa can have 80 percent of their population living on farms, but because of low productivity and poor infrastructure they can still be the least food-secure places in the world.

"It's simply not practical to turn all our farming into a Michael Pollan–style idyllic agriculture," Carrington said. "Especially in developing regions of the world like sub-Saharan Africa and elsewhere, food security crops like cassava, millet, and cowpeas have been produced in ways that would meet organic production standards. But the lack of tools, technology, transportation—all the result of a lot of different things—means that organic production has been exceptionally unproductive. It's a horrible, unfortunate situation.

"Our research aims are to improve the sustainability of agriculture, to improve the strength of plants to do as much of the work as possible, as opposed to using insecticides and fungicides," Carrington said. "These are not organic versus conventional issues. These are universal issues. We just happen to include GMO in the toolbox. It is not a panacea, and it is not the only tool in the toolbox. It is just one of dozens. But removing that tool we do at our own peril. We don't gain anything; we actually lose the ability to solve a lot of important problems that affect real people."

Michael Hansen, of the Consumers Union, has been hearing such arguments for two decades. Back in June 2000, around the time golden rice was on the cover of *Time*, Hansen went to a congressional hearing to hear what people were saying about the technology's prospects.

"Right there on the invitation was the statement that golden rice was already saving the eyesight of thousands of children in Asia, and that was all false," he told me. "Fourteen years later, it hasn't saved the eyesight of a single child."

If you go back to the early 1980s, when genetic engineering was in its infancy, everyone said GMOs could do all these wonderful things, Hansen said. Fast-forward to today, and genetic engineering has all been about herbicide tolerance. By now, some 94 percent of our soy is GE, along with 99 percent of our sugar beets, 93 percent of our corn, and 92 percent of our canola. All of these crops were designed to be herbicide tolerant.

"All those original claims really only led to an explosion of glyphosate, so they needed something to show that GE will benefit people," Hansen said. "Golden rice is still being used to get good PR for industry, to show that they can do something that is clearly good. But look at the millions of dollars that have gone into golden rice, and contrast that with what they've done in the Philippines with the more traditional things they do. They give little pills to people as vitamin supplements, at a cost of about twenty cents a person per year. They have been fortifying noodles with vitamin A. All this was done with no money poured into it."

Hansen would like to see considerably more investment in traditional foods that support local growers rather than global conglomerates, and that have evolved over millennia to endure droughts and pests. In some places, this is happening. In Kenya, farmers have recently increased the area planted with such indigenous greens (rich in protein, vitamins, iron, and other nutrients) by 25 percent.

Yet in June 2016, more than one hundred Nobel laureates signed a letter urging Greenpeace to drop its opposition to golden rice. "We urge Greenpeace and its supporters to re-examine the experience of farmers and consumers worldwide with crops and foods improved through biotechnology, recognize the findings of authoritative sci-entific bodies and regulatory agencies, and abandon their campaign against 'GMOs' in general and Golden Rice in particular," the let-ter stated.

Greenpeace refused to back down. "Accusations that anyone is blocking genetically engineered 'Golden' rice are false," the group said. "'Golden' rice has failed as a solution and isn't currently avail-able for sale, even after more than 20 years of research. As admitted by the International Rice Research Institute, it has not been proven to actually address Vitamin A Deficiency. So to be clear, we are talking about something that doesn't even exist.

"Corporations are overhyping 'Golden' Rice to pave the way for global approval of other more profitable genetically engineered crops. This costly experiment has failed to produce results for the last 20 years and diverted attention from methods that already work. Rather than invest in this overpriced public relations exercise, we need to address malnutrition through a more diverse diet, equitable access to food and eco-agriculture."

Michael Hansen, who was trained as an evolutionary biologist, spends a great deal of time traveling around Asia teaching farmers and consumers about farming and food. Recently, in the Philip-pines, he saw an ad on the side of a bus for chicken tenders that had been fortified with vitamin A the old-fashioned way—and not with GMOs. In the Philippines at least, vitamin A deficiency has plum-meted, and it has had nothing to do with golden rice. "If you want to get at these problems, you can deal with the symptoms or with the causes," Hansen said. "This is basically an issue of poverty. All poor people can afford to buy is rice. So the way to get people out of that, you have to deal with poverty. What if all that money had been put into food-fortification techniques, or to teaching people to grow foods that are high in beta-carotene that people actually eat? Mangoes, yellow maize, papaya, yams, red peppers, spinach, cabbage—all have high levels of beta-carotene, and these are all foods that are culturally appropriate.

"Golden rice is just a PR move," he said. "Now they want to do golden bananas, engineered with vitamin A, and they want to do trials in Uganda. Nobody has published data that has shown these are safe. Yet they are already doing some kind of feeding trials. They're talking about golden corn."

Alfred Sommer, the vitamin A expert at Johns Hopkins, takes a different view. To his mind, GMO critics oppose golden rice because they fear it would be a smashing success, and thus ruin their chances of opposing GMOs elsewhere.

"I've come to accept that GMOs are in fact just a more sophisticated version of hybridization," Sommer told me. "This is just more rapid than crossbreeding things. It's not like there's a mad scientist out there trying to grow people out of corn.

"I can understand people being concerned, but the fact is that all our soy is GMO, and nobody seems to have been hurt by that," Sommer said. "Big Ag will do what Big Ag does. Today, nobody notices that soy is all GMO, and in twenty years nobody will remember this."

The Plant That Started Civilization, and the Plant That Could Save It

f you think about the GM grains that prop up the world's global food system, there is one leg in the stool that is mysteriously missing: wheat. How have corn and soybeans become almost entirely GM, and wheat—the plant that started civilization 10,000 years ago, and that covers tens of millions of acres in the United States alone—has not?

It has not been for a lack of desire. Monsanto began testing Roundup Ready wheat on experimental farms back in 1994, and in 2004, the FDA declared that the company's wheat posed no health or safety risks. But that same year, Monsanto abandoned plans to release its seeds onto the market, a dramatic decision that food safety advocates (who had long complained that the FDA's information had been provided by Monsanto and did not include the FDA's own tests) considered a "watershed event."

What happened? Why aren't we all eating GM bread, and bagels, and hamburger buns?

One reason has to do with the unimaginably intertwined global food system, which (when it works) can get grain grown on one continent to food processors on another continent to consumers on a third continent. When the system doesn't work, when something gets into the system that is unexpected, unapproved, and unwanted, the whole thing can come to a screeching halt.

Just ask Larry Bohlen. In the summer of 2000, while working for the environmental group Friends of the Earth, Bohlen was spending his time looking over the EPA's approval process for GM crops. One day, he noticed something strange about the approval of a strain of GM corn called StarLink. The corn had been engineered by the company Aventis to carry the insecticide *Bacillus thuringiensis*, or Bt, which kills the destructive European corn borer. Because the corn contained a protein called Cry9C, which the EPA suspected might cause allergies in people, StarLink had not been approved for human consumption, but Aventis had still been allowed to sell Star-Link to farmers to grow corn for animal feed.

Given the complexity of the crop and food distribution systems, this seemed absurd to Bohlen. "We were in conversations with farmers who were telling us that most farmers do not separate genetically engineered corn from conventional corn," Bohlen said. "Given that very little of the corn is separated and there's a type of corn not approved for human consumption, I thought there was a good chance that it had made it into our food."

Bohlen decided to check for himself. He went to his local Safeway and bought twenty-three different corn-based products: boxes of cornflakes and taco shells, tortilla chips, a corn muffin mix, some cornmeal, a couple of enchilada TV dinners. He shipped them to a laboratory with a simple question: Did any of the products contain the unapproved protein Cry9C? If any did, it meant that an entire stream of the country's river of processed food might have been unintentionally contaminated by GM corn.

The results, in one case, came back positive: the lab found Star-Link corn in taco shells branded by Taco Bell. All taco shells containing StarLink corn were recalled.

Inside the food industry, the revelation caused an explosion: this was the first GM food to be recalled nationally. Suddenly, food processors could no longer be sure the grain they were buying was free of unapproved GMOs. Although scientists had long assured them that GMOs were safe, the reaction was very much a fear of GM "contamination." If consumers decided they wouldn't eat foods with GM ingredients, it wouldn't just be Monsanto (and the $20 billion biotech industry) that would suffer; it would be any company (in the $500 billion food industry) that made food from Monsanto's grain. If these two industries diverged, Bohlen told me, "biotech would shift from being a growth industry to being a struggling commodity industry."

For consumers, the StarLink story offered a rare (and bewildering) look inside the mysterious system that delivers Americans processed food. The corn, it turned out, had come from farmers in six different states, who had shipped their grain to a miller in Texas, who had ordered conventional corn and (unwittingly) gotten GM corn instead. The cornmeal was then sent to Mexico, where it was processed into taco shells, then returned to the United States to be distributed everywhere by Kraft Foods.

Once the news broke in September 2000, food that had been contaminated by StarLink corn started popping up all over the place—and not just in the United States. It was found in Japan, Korea, the United Kingdom, and Denmark. Aventis officials said they had "difficulty imagining how our corn could end up in the human food supply."

To Larry Bohlen, it was entirely obvious. "Aventis made a big mistake by assuming that thousands of people making decisions every day on their farms would be able to separate the StarLink corn from conventional corn," Bohlen said. "Harvest days last for fourteen hours. Farmers are driving late into the night. They're under a lot of pressure. Farm prices are really low. There's even pressure for some people to sell the StarLink into the food system to get a higher price. There are so many reasons that the StarLink corn can get into the food supply that it was a risk that wasn't worth taking."

Bohlen also considered the StarLink debacle evidence of the gaping loophole in the food-testing process. "We've been saying for a long time that federal authorities should be doing this testing, but so far it's been left to groups like us," he said. Aventis ended up spending $500 million to withdraw StarLink from the corn market. But it proved much harder to undo the anxieties the contamination had caused—among food manufacturers as well as consumers. Where once the food industry had been in lockstep with the biotech industry, the StarLink affair proved just how precarious this marriage could be.

Things have only gotten rockier with the development of GM "biopharm" crops, which pharmaceutical companies are developing to create drugs. Though the USDA has approved more than 300 biopharm plantings around the country since 1995, both states and traditional farmers—worried about StarLink-style contamination— have been more suspect. In 2005, Arkansas-based Riceland Food, the world's largest rice miller, asked federal regulators to deny a permit to a company hoping to plant GM rice for the manufacture of an antidiarrheal drug. Anheuser-Busch, the country's top buyer of rice (as well as its largest brewer), said it would no longer buy any of the $100 million worth of rice grown in Missouri if GM rice was allowed to be grown anywhere in the state.

California recently rejected a company proposal to grow rice engineered with human genes after traditional rice growers said even the *prospect* of contamination would scare off international markets. In 2011, the German conglomerate Bayer (which now owns Aventis, the same company that made StarLink corn) agreed to pay $750 million in settlements to 10,000 farmers who claimed the company's GM Liberty Link rice contaminated their domestic crops and drove prices down; global markets just weren't buying it. "What has really alarmed the food industry was the idea that they might get corn in their cornflakes that had someone else's prescription drugs in it—either by getting mixed up or through cross-pollination," Bohlen told me.

It was right in the middle of all this that Monsanto—despite spending a decade designing it, and despite FDA approval—decided not to release GM wheat. The company still maintained a handful of experimental wheat plots, though, and in 2013, something strange happened. A farmer in Oregon—trying to clear his field by spraying Roundup—found he couldn't kill his own wheat. Some of Monsanto's experimental GM wheat seeds had somehow made their way into his fields.

Suddenly, it was StarLink all over again. As the news of this contamination spread, Japan and South Korea immediately suspended U.S. wheat imports. European officials urged greater screening of U.S. grain. Lawyers for American farmers threatened to sue the company. And—despite USDA assurances that American wheat remains GMO-free—the global prospects for GM wheat were once again put on hold.

And for a team of world-renowned scientists working in the middle of Kansas wheat country, that was just fine.

Reinventing the Plains

Salina, Kansas, sits on the western edge of a thousand miles of corn and soy and wheat. Down at the end of Water Well Road, where the pavement runs out, a small group of plant researchers is leading the effort to overthrow the entire American agricultural system. The Land Institute looks nothing like the Danforth Center. There is no glass and steel here, no ornamental fountains, no multimillion-dollar laboratories. And there is no evidence of corporate agriculture.

Inside the president's office, a ramshackle affair with overstuffed wooden bookshelves and a small refrigerator filled with good beer, an eighty-year-old bull of a man named Wes Jackson is holding forth about GMOs, pesticides, and ridding the world of problems caused by "that outfit in St. Louis."

By which he means Monsanto.

"The idea of Manifest Destiny—of wiping out the Indians and going to the moon and building a supercollider—it seems like humanity doesn't have the capacity to practice *restraint*," Jackson told me. "So you say you can feed 7 billion people, then you can feed 9 billion, then what? A woman recently said to me, 'What about all these new planets?' I said, 'Good: buy one-way tickets, and pay for it yourself.' People have these escape clauses, and that's just being a dummy."

Jackson is one of the most influential thinkers on agriculture in the country, a renowned scientist with a vision for changing the face of American farming in dramatic, even radical ways. He is also a master of rhetorical flourish and pushes his vision with the stentorian voice (and the moral urgency) of a preacher. The way we grow our food is part of a much larger problem in the way we treat our land, our water, and our climate.

"There are too many of us, but our consumption is rapacious," he told students graduating from the University of Kansas in 2013. "It is legal to rip the tops off mountains, get the coal and burn it. It is legal to drill for oil and natural gas—from the Gulf to the Arctic—and burn it. It is legal to engage in fracking that threatens groundwater to get natural gas and burn it. It is legal to have our soils erode and toxic chemicals applied, legal to allow our rural communities to decline and watch so much of our cultural seed stock disappear."

Jackson places much of the blame for this state of affairs on what he calls the "industrial hero," the scientist (or more broadly, the corporation) claiming to have high-tech, silver bullet answers to highly complex problems.

"These technology fundamentalists are far worse than religious fundamentalists," Jackson told me. "The ultimate fundamentalist doesn't even know he's a fundamentalist. If your efforts are clouded by the desire for financial gain, or clouded by a desire to be famous, then you are not available for the pursuit of wisdom. Where does responsibility come from? 'Feed the world' is a very poor veil to put around the desire to get rich or famous. Rather than think hard about problems, the industrial hero says, 'We must feed the world,' which has an easy move to a profit agenda. Contrast that with the phrase 'The world must be fed,' which carries a social agenda that has to do with social justice. You have two different breeds of cat."

Mention publicity-generating moves like Monsanto's recent announcement that it would contribute $4 million to study the decline of the monarch butterfly, and you can practically see Jackson's blood start to rise. Sure, the money is nice, he says, "meanwhile 97 million acres of corn is going to be drenched with Roundup. That's the problem we allow ourselves."

Jackson has been thinking about feeding the world for five decades. He has published highly influential books, built an internationally recognized research station on a shoestring budget, and been showered with honors (including a MacArthur "genius grant") and recognition (he was named by both *Life* magazine and the Smithsonian as one of the twentieth century's most influential people). All this for work he and his research team have been doing on a plant that has yet to reach the market. When it does, Jackson hopes, farmers may be able to grow vast amounts of food and begin repairing land that has been degraded for centuries.

One of the first things Jackson likes to show visitors is a pair of vertical posters, hanging above a stairwell in the institute's lab facility. To the left is a picture of a wheat plant, its thready roots extending just inches below the soil. To the right is a picture of a plant Jackson and his scientists have been developing for many years, something known as intermediate wheatgrass. Its roots drop down fully ten feet.

"The picture to the left is the plant that started civilization," Jackson told me. "The one on the right is the plant that's going to save it."

The Problem with Annuals

Although he is one of the country's leading thinkers about agriculture, Wes Jackson is not particularly interested in what's going on in the vegetable aisle. He wants to change the rest of the grocery store. Why? Because 70 percent of the calories we eat—and 70 percent of our farmland—is wrapped up not in fruits and vegetables but in grains. So are 70 percent of the soil erosion and 70 percent of the petrochemicals.

"You have all these showy images of green, leafy healthy stuff, but humans are grass-seed eaters first and legume-seed eaters second, and the rest is water," Jackson said.

For a farmer and a plant scientist, Jackson is rather cantankerous about the way people plant the land. He doesn't like to talk about problems in agriculture. He likes to talk about the problem *of* agriculture—a problem that has gotten more complex (and dangerous) since people first put plow to ground.

This is the story he tells. People have been sowing, reaping, selecting, and trading seeds since they first started growing food 10,000 years ago. If a certain plant did well in a certain soil in a certain climate, the farmer would save the seeds and plant the cultivated variety, or cultivar, again the next year. If a plant did poorly, it was discarded. Human selection mirrored natural selection, but with a twist: the only plants that got to pass on their genes were those that proved useful for the human appetite.

The trouble began when early farmers chose to breed plants that needed to be planted every year, rather than plants that remained in the ground for years at a time. Although the vast majority of plants on earth are perennials, annuals are far more easily manipulated by farmers (and later by genetic engineers). By choosing seeds from only the hardiest or most productive plants, Neolithic farmers (and many generations that followed) could develop crops that yielded more and more food with each passing year. Go to southern Mexico, the birthplace of corn, and you will see a vast variety of corn species: some are two feet high, some are fifteen. Some are blue, some are yellow. All of these crops were selected, over many generations, to achieve flavors and textures and hardiness unique to the specific places and cultures in which they are planted.

As breeding science got more and more sophisticated, farmers learned to speed up this process by crossbreeding (or hybridizing) annuals. These crops have an obvious advantage for growers: because they live only a year, they can push a higher percentage of their energy into producing big, high-calorie seeds. (Genetic engineers are correct when they say they have simply taken this one step further, configuring annual corn and soy to resist herbicides, or drought, or insect infestations.) Because all our mountains of annual corn, wheat, and soy can be ground up and turned into everything from breakfast cereal to cattle feed to soda pop, the food industry has turned these crops (for better or worse) into the staples of our national food system.

But all this efficiency, all this uniformity, has come at a significant cost.

Jared Diamond, the scientist and bestselling author of books like *Guns, Germs, and Steel*, has called annual agriculture "a catastrophe from which we have never recovered." If all of human history is represented by a twenty-four-hour clock, Diamond writes, humans were hunter-gatherers from midnight until 11:54 p.m. We've been *farmers*—let alone genetic engineers—for only about six minutes. Not a long time, really, to get things right.

Long before GMOs, farmers figured out how to grow lots of cheap calories. But this often came at the expense of poor nutrition, a fact that has become vastly exaggerated in our industrial food era. African Bushmen eat some seventy-five different wild plants, Diamond writes, and could never die of starvation in numbers like the Irish did during the potato famine because of the variety in their diets. The Irish were so dependent on this single crop that when, in 1845, a fungal blight knocked out 90 percent of the crop, a million people—fully one-eighth of the country's population—starved to death. Another 1.3 million people left the country, followed by 5 million more over the next several decades.

Indeed, it has been argued that far from relieving famine, farming actually *contributes* to it. When a culture's food supply becomes centralized—dependent on a few crops, grown either by a few companies or by a central government—bad things tend to happen. In the late 1950s, during China's Great Leap Forward, Mao dictated that both wheat and rice be planted at densities far beyond the soil's capacity to support them. The result was disastrous crop loss, and some 80 million people died of starvation. In the Soviet Union, Stalin subdued Ukraine (and killed 7 million people) by controlling Europe's breadbasket.

So consider that the United States' 320 million acres of farmland is planted with 80 percent annual, monoculture crops. (The other 20 percent is perennial—mostly hay and alfalfa for animal feed.) Although these plants are perfectly designed to fit the industrial economy that designed them—perfect for processed food; perfect for ethanol; perfect for feeding 10 billion cows, chickens, and pigs— they are also freighted with serious drawbacks. Annual plants have very shallow root systems that tap into only the top few inches of soil. Add to this annual plowing and spraying, and the ground quickly becomes nutritionally depleted, even barren. Tim Crews, a Land Institute scientist, considers the soil beneath traditional monoculture crops to be "just this side of a Walmart parking lot in terms of ecological health."

Depleted soils, of course, force farmers to use enormous quantities of synthetic fertilizers, which are derived from petrochemicals. But since only about half of these fertilizers actually gets absorbed by plants, the rest ends up elsewhere. Some gets converted into nitrous oxide, a potent greenhouse gas. Some washes downstream where (if we're talking about the Midwest) it eventually ends up in the Gulf of Mexico, where it creates algae blooms. When these massive pulses of aquatic plant growth die back and decompose, they suck so much oxygen from the water that they leave dead zones that can be seen from outer space.

Because annual plants must be stripped out and replanted every year, the soil beneath them requires vast volumes of herbicides to control the weeds that sweep onto the bare soil. The bare ground is also terribly prone to soil erosion. Even though modern no-till farming has reduced soil loss by 40 percent since the 1980s, many farmers are still tilling extensively, and we are still losing some 1.7 billion tons of topsoil every year. This is a lot when you consider that, left to itself, soil replenishes itself at a rate of about a quarter inch per century. We're running things down in a few decades that took millennia to create.

It is for these reasons that Wes Jackson considers the GMO question to be something of a bait and switch. From the beginning, biotech companies have said they were developing technologies to "feed the world." What they have in fact done is use the technology to ramp up an already destructive form of industrial farming.

Stan Cox, a plant breeder at The Land Institute, likes to show people a sheet with two columns on it: problems caused by conventional farming in 1990, and problems caused by conventional farming in 2015. The lists are exactly the same: soil loss and degradation, toxic chemicals, water pollution, monoculture, factory farming, corporate control. Except for one thing: the 2015 list includes GMOs.

In other words, even as we squeeze an incredible amount of food energy out of limited cropland, we are pushing plants, soil, and the environment as a whole to their ecological limits. And even if we get rid of GMOs, in other words, we would still have all the problems caused by large-scale farming. It's not the *technology* that causes the real problems. It's the *system*.

"Let's take the big picture for a moment," Jackson told me. "This

GMO thing prevents real thought. It's worse than a digression. It's a *distraction*. We are really interested in social justice and reducing greenhouse gases and reducing poverty. How do we meet a bona fide human need—like reducing poverty as we reduce fossil fuel use—how do we meet that need in a way that goes beyond GMOs? This pipsqueak thing of GMOs enters into the arena and tries to create balloons that fill the whole arena—that's the problem. They de facto baited us to be pulled into a distraction, and we've taken the bait. Our role is to make the subject as complicated as it actually is."

The Promise of Perennials

When he and his family started The Land Institute in 1976, Jackson saw in the Kansas landscape both a great problem and a great solution. By the 1970s, despite great federal conservation efforts and federal financing, the soil on monoculture farms was still eroding at about the same rate it had during the dust bowl of the 1930s. By dramatic contrast, Kansas's native prairies—the few that had not been plowed under for crops—were gems of sustainable growth.

Instead of the ecological desert of the monoculture farm, prairies were "perennial polycultures": gorgeous, diverse tapestries of soil, plants, insects, animals, and birds that worked in a rich balance that kept the entire system healthy. Prairies did not require annual planting because the plants were virtually all perennials. They didn't require excess petrochemical fertilizers—they just soaked up energy from the sun and nutrients from the soil. Plants that needed soil nitrogen, like wild grasses, were helped by legumes that provided it, like bundleflower. Prairies are not generally plagued by weeds, because their perennial leaf canopies and roots outcompete invasive species. Prairie soils do not wash away in the rain, because perennial

plants have far more extensive and woven year-round root structures. Because they have evolved to survive for multiple years, they
have developed better disease resistance and can withstand stress
(like drought). And because they have evolved diverse relationships
with both plants and animals around them, prairie perennials tend
to survive (as a system) even if a single species declines.

So Jackson had an idea: Why couldn't a farm—or even an entire
country's farmland—function more like a prairie? Would agricultural "biomimicry" work?

The idea seemed so obvious, Jackson couldn't believe no one
had tried it before. If farms—even large-scale farms—mirrored
prairies, they would solve a wide variety of intractable problems in
industrial agriculture. Because they have permanent root systems,
they could eliminate more than half the soil erosion in the United
States, saving $9 billion worth of fuel for tilling equipment every
year. They would also save nearly $20 billion worth of soil—though
how one puts a price on an essential, nonrenewable resource is a bit
of a parlor game.

Deep roots would also mean the efficient use of both water and
soil nutrients, especially nitrogen—which could radically cut back
on the need for both irrigation and fossil fuels, especially synthetic
fertilizers. Because perennials outcompete weeds, they would not
require herbicides. Because they would not have to be torn up and
planted every year, they would improve the land's biodiversity, both
underground (in terms of microorganisms in the soil) and abo-
veground (in terms of food and habitat for everything from bees and
monarch butterflies to migratory birds).

With so many obvious advantages, why hadn't farmers been
planting perennials for 10,000 years? To Jackson's scientific mind,
the answer lay in the way perennials are built. Because perennials
must store energy in their roots to survive year after year, they can-

not afford to put all their energy into producing seeds, as annuals do. On the other hand, perennials make up for some of this because they enjoy a much longer growing season.

So for Jackson's team at The Land Institute, the trick has been trying to figure out how to take the best traits from a perennial plant (deep roots) and combine them with the best traits of an annual (big seeds). How do you take a 10-foot-tall perennial grass and domesticate it to create bigger seeds and higher yields? How do you get wild plants to accept uniform planting and efficient harvesting?

The work has been slow and laborious. Jackson's scientists, who have been working on this for thirty years, say their work is "like scratching off lottery tickets." But if they can figure it out, they might actually alter the course of agriculture that has been in place for 10,000 years.

Perennial Calories

One of the perennial plants Jackson and his team are betting on is a wild relative of wheat from Persia known as intermediate wheatgrass, which produces a (trademarked) grain called Kernza. If it strikes you as odd that the savior of American agriculture might come from Iran, consider that all American wheat—all those "amber waves of grain"—comes from Central Asia and the Middle East, where it has been grown for at least 7,000 years.

In the early 1900s, immigrant farmers like Central European Mennonites brought wheat seed and planted it across the middle and upper Midwest, initially for cattle forage. Now, the Midwest is blanketed with more than 60 million acres of wheat: hard winter wheat blankets Kansas, Nebraska, Oklahoma, and North Texas;

spring wheat covers North and South Dakota, and eastern and central Montana.

Replacing 60 million acres of annual grain with perennial grains would be an ecological coup of historic proportions. But this is hardly the limit of Jackson's imagination. Intermediate wheatgrass can also be used as a biofuel, which means it could replace some of the 40 million acres of corn American farmers currently grow not for food but for ethanol. (Claims from industrial corn companies that ethanol is a "green" substitute for petroleum have long since been discredited; they offer virtually no benefit in terms of reducing carbon emissions. And consider that Brazil is cutting down a million acres of rainforest a year to plant biofuel corn, then ships half of this fuel to Europe. The net effect of this transfer is 50 percent more carbon emissions than gasoline.)

So, just for starters (and not even counting the perennial rice the Land Institute's colleagues are developing in Asia or the perennial sorghum they are working on in the United States and Africa, or the perennial silphium, a sunflower relative that can be used for oilseed), that's 100 million acres of Walmart parking lot that could be turned into highly productive, ecologically diverse, carbon-sinking perennial polyculture.

Given the state of our food system, and the state of the climate, the stakes are high. "The world is not going to be fine waiting around for thirty more years for crops that could be sequestering carbon," Tim Crews told me. "It's not as though there's nothing to lose here."

With world-changing ambitions resting on just a couple of plants, the trick for scientists like Crews and his colleague Shuwen Wang is to take a wild plant (like intermediate wheatgrass) and persuade it to mate with a domestic plant (like wheat) to produce food and fuel that people will like so much they'll be willing to shift their entire approach to the way their land is planted. As is, the grain can be ground up and mixed with regular flour for things that don't need a big rise, like pancakes or cookies. It can be used to make beer. But so far intermediate wheatgrass produces less than half the yield of wheat and (left to itself) doesn't create enough gluten to make bread. One gold ring for The Land Institute's plant scientists is clearly a marriage between the perennial hardiness of intermediate wheatgrass and the big seeds, high yields, and gluten of wheat.

Just as it does for genetic engineers, the secret to this marriage lies in the mysteries of genetics; the team at the Land Institute can't wait 7,000 years (the time it took domestic wheat) for intermediate wheatgrass to produce big, plentiful seeds that release easily during threshing.

"There's still a lot of things we don't know yet, and yields and seed size are still not where we want them," Lee DeHaan told me. "But we thought it would take a hundred years to get domestication, and we have been surprised at how fast it's been coming along."

DeHaan, who grew up on a farm in Minnesota, came to The Land Institute in 2001, armed with a PhD in agronomy and agro-ecology. He and his team are using two approaches to solve the domestication problem. The first is hybridizing wheat and intermediate wheatgrass to genetically nudge their offspring to express the best traits of both. The second is growing (and selectively breeding) enough generations of wild wheatgrass to get it to perform more like wheat. The goal is essentially the same: Create a plant that makes big, plentiful, easily harvested seeds and also comes equipped with deep roots—and will come back year after year. They are doing this work without engineering any genes, which, especially compared with the relative glitz of engineering powerhouses like the Danforth Center, makes work at the Land Institute both slow and "unglamorous," DeHaan said.

"It's hard to attract scholars, students, and funding. It looks so tedious and boring," he said. "There's something about lab work that seems so much more attractive and prestigious. The excitement always is for what's new. Funders love that because it's always about what we couldn't do in the past, now we can do. You have all these expensive machines, data being calculated on computers—all that versus slogging it out in the field."

While DeHaan remains suspicious of the industrial-scale ends to which most GM technology is used, his goal is sustainability, not purity. He's looking for clues to what makes a domesticated plant's seeds nonshattering (meaning the seeds stay in the plant's seed head, rather than blow to the winds, as they do in wild plants) and also free threshing, which means they are easily separated from the chaff during harvesting. His work has long been fully engaged with the techniques of molecular biology—gene sequencing, molecular marking, chromosome staining—that help him figure out a plant's genetic structure. He's just not taking genes from one plant and putting them in another.

"If DNA is a book, we just want to read it. We're not cutting pages from other books and inserting them into this book," De-Haan said. "Without molecular tools, perennial wheat will never be a reality. If you find a unique marker associated with something like free threshing, you can generate thousands of different segments and see where they lead. This is not GMO, just the ability to se-quence genes, to stain chromosomes, track molecular markers, fig-ure out the function of genes. Because these tools are relatively cheap, it doesn't make sense not to use them."

Besides his work with wheatgrass, DeHaan and his Land Institute colleagues are also working with scientists in China to develop perennial rice, and with scientists from Africa on perennial sorghum. The day of my visit, he was pawing through a patch of sorghum plants, their seed heads drawn together inside slender paper bags. DeHaan was rather urgently asking the plants to mate.

Sorghum looks and acts a bit like corn: it is a large grass plant, with energy-rich seeds that can be used for food, forage, or fuel. Like corn, it has also become one of the world's most important crops; the world currently produces about 70 million tons a year, ranking behind only wheat, corn, barley, and rice. In the United States, 7 million acres of sorghum are planted yearly, mostly for animal feed and to produce biofuels. In Africa, however, sorghum is a critical source of calories on a continent with exhausted soil and the constant threat of drought. After a few weeks without rain, corn will shrivel up and collapse. Sorghum will just stand there and take it. When the rains return, sorghum will start growing again.

At a nearby lab table, Pheonah Nabukalu, a Land Institute scientist from Uganda, was measuring out sorghum seeds on a small scale and entering data into a computer. Around her, on the table, on nearby racks, were scores of brown paper lunch bags filled with seeds. She was poring over some 500 different experimental lines, all bred in temperate conditions, trying to decipher which plants have the highest yields, and selecting those that have perennial rhizomes. Even if she finds a promising candidate, there remains the challenge of getting a temperate-raised seed to grow in a tropical place like Uganda—or the other places she thinks might benefit, like Mali, Ethiopia, and South Africa.

Armed with a plant breeding PhD from Louisiana State, Nabukalu has been working on developing high-yield grains that are also resistant to pests, diseases, and drought. One project involves American seeds that have already been crossed with perennials, and then crossing these with varieties native to Africa. Using local seed lines has its benefits, since local seeds have already been bred to resist local pests and diseases. But regional specificity is also critical for developing crops that will fit well into the local environment. At field stations in Uganda, she is helping oversee experimental crops in arid, semi-arid, and rainy locations. And she is doing this work slowly, and without GM technology or corporate influence.

"Companies always want to do it fast because of the money," she said. "The seed companies working in Uganda are mainly working with Monsanto and with corn. Corn is not even native to Africa. It's a colonial import from Mexico, and it's only been in Africa for a hundred years."

She is also suspicious of GM cassava—the kind Nigel Taylor is growing at the Danforth Center—but not for the usual reasons.

"I've never seen orange cassava," she said. "Our sweet potato is also white fleshed. We cook it for three to four hours, but it holds up nicely. When you cook orange flesh for three hours, it breaks down. Making farmers switch is very hard. Taste is tradition."

Could Perennials Be GMO?

DeHaan's and Nabukalu's ambivalence about biotechnology—interested in some techniques, suspicious of others—is shared by other scientists at The Land Institute. At one level, they can see the benefits: surely it might be easier (and faster) to nudge wheatgrass toward bigger seeds by stitching in a few genes from a wheat plant. And what if you could engineer a plant (like corn) to fix its own nitrogen, like a legume? Think how much petroleum-based fertilizers would no longer have to be applied to tens of millions of acres of nitrogen-fixing corn.

"The anti-GMO side is too fearful," Wes Jackson told me. "It reminds me of something kids used to say on the playground: 'When in worry, when in doubt, run in circles, scream and shout.' It's okay with me to look at sequenced gene segments, to help speed up the research process. The term 'GMO' is a *generic*—to come out against them is not to consider the *specifics*."

The anti-GMO movement "sucks way too much bandwidth away from many other aspects of sustainable agriculture," Tim Crews said. "If we somehow got rid of all GMOs—if sustainable agriculture somehow reared up and achieved this Herculean feat— we would still be back in 1990 with the same list of profound short-comings in agriculture."

Since GM research is so capital intensive (done almost exclu-sively in corporate laboratories, or in university labs funded by cor-porations), it has become almost entirely focused on what Crews calls "patentable objectives": crops that can make companies a lot of money.

And it's this dynamic—corporations getting their hands on technology, and then scaling it up—that causes deep and unpredict-able problems.

By this reckoning, GMOs are like pharmaceutical drugs: they are so expensive to design, and test, and market, that only the big-gest corporations have the wherewithal to introduce them. Because of this capital investment, corporations will pursue only those crops that offer a return on their dollar.

But GMOs are like pharmaceuticals in another way as well. In the right hands, opioid pain pills, for example, can work a kind of therapeutic magic. They have made pharmaceutical companies billions of dollars. But as they have flooded into popular use, pain pills have also caused epidemics of addiction, black markets, and misery.

This is precisely the kind of "cascading" consequence that worries Crews about GMOs. Let's say scientists can figure out how to increase a plant's photosynthetic efficiency. What could go wrong by increasing a plant's energy production? Wouldn't that just make food crops that much more productive?

"If they could raise that ceiling—that's the Holy Grail," Crews said. "So then what? All of the other resources that are synched up in that ecosystem—all of those things will get out of whack."

Taking his cue from the Italian researcher Mario Giampietro, Crews offers a dark example: Consider a spider that can suddenly make a bigger web to catch more insects. Suddenly all spiders are doing this, making bigger webs and catching more insects, until the population of the insects crashes, which leads to a crash in spiders as well. Or consider the burning of fossil fuels: great power, great convenience—and then the countless cascading effects of climate change. Entire natural systems get thrown out of balance.

"Right now, we have ecosystems that have evolved to be in sync with resources that are available through natural processes over the course of a year," Crews said. "Cacti are in sync with the rain available in a desert, redwoods are in sync with the rain on the coast. When you tweak a plant's genes to make them more productive, you can stress the larger processes—both the plant and its surrounding ecosystem.

"Let's say you take a grass like oats and make it more photosynthetically efficient. All of a sudden the plant needs all its nutrients in much greater quantities. Which leads to a dynamic like we had during the Green Revolution: all this new productivity requires a huge new increase in nitrogen or phosphorous fertilizer. All of a sudden these nutrients—especially phosphorous—become taxed, or even tapped out, or, if we're talking about nitrogen, require a massive increase in petrochemicals.

"Then, if those genes get out into the world through cross-pollination with wild plants, the same would be true for natural systems," Crews said. "If the resources are there, these plants would immediately take over, becoming taller, using more resources, out-competing other plants. Or you run into situations where an ecosystem that is already resource limited grows beyond what they had been before, and what that looks like I don't even know, but it would be a novel situation."

Like Lee DeHaan, Crews is not dead-set against the use of GM technology. If GMOs could move genes between plants that already cross naturally (or cisgenically), "why would we not go there?" Crews said.

"There could be a marginal improvement with GMO traits, but it's still the wrong approach for addressing agriculture's shortcomings. The overall sociological phenomenon is not addressing agriculture in ways that need to be addressed. If you take an ecological approach that solves things on this list, then we're talking, whether it involves conventional breeding, or cisgenics, to develop ecological agriculture.

"The way we're doing things is long term and messy," Crews said. "We are by no means simply trying to introduce a new species here or there to be the next superfood or fill some niche market. Our rather long-term ambition is to replace tilled agriculture with something that is far more ecologically complex and sustainable—agriculture ecosystems that humans could be proud of rather than the most compromised ecosystems imaginable."

Can GMOs Be Sustainable?

No matter where you stand on GMOs, it seems reasonable to ask if our food system hasn't somehow become too big to fail. If industrial farming is the real culprit in our national eating disorder, perhaps a solution can be found simply by scaling back. Not in the interest of purity, or in pursuit of utopian visions of food. Not to get rid of GMOs, or even to (completely) rid the world of pesticides. Perhaps it's enough to have food produced on a smaller scale, by farmers who take excellent care of their land.

To get to Jennie Schmidt's farm, you drive east across the Chesapeake Bay Bridge, then hook a left and head north through an endless tapestry of some of the most fertile land on the Eastern Seaboard. Her Maryland farm is just beyond a shooting range, out along Sudlersville Cemetery Road. When I arrived on a cold February day, the Schmidts' dog Dozer met me out front. Jennie apologized for the carpet guys who were replacing the wall-to-wall carpeting that had been covering the floors of their modest home for thirty-seven years.

The Schmidts work 2,000 acres. It's not a huge operation, by

Iowa standards, but consider this: Last year the Schmidts produced 12 million pounds of Roma tomatoes, enough to fill twenty tractor-trailers every day for two weeks. Over the years, the Schmidts have experimented with sweet corn, sweet peas, and lima beans, but given the risks—the Eastern Shore gets forty-five inches of rain a year, much of it during spring growing season—they have settled on their current mix: Tomatoes and green beans for the vegetable market. Soft red wheat, which they ship to a processor in Pennsylvania to make into crackers and pretzels. Twenty-two acres of grapes for local vineyards.

And 1,500 acres of soybeans and corn, a good deal of it grown from GM seeds.

As third-generation farmers, the Schmidts have found a sweet spot between small-scale farms that survive by supplying farmers' markets, and industrial-scale operations that must invest millions into their own harvesting and canning infrastructure. Jennie Schmidt calls her farm a "supermarket farm," but not because most of her crops go into the vegetable aisle. They don't. She and her husband, Hans, grow for canneries, which turn their tomatoes into "value added" products like salsa and tomato sauce. Tomatoes harvested at her farm in August are trucked to Pennsylvania and are in jars or cans less than forty-eight hours after they are picked.

Sitting as it does in the middle of the Eastern Shore—a lobe of beautifully fertile land stretching from near Wilmington, Delaware, all the way to the bridge to Virginia Beach—the Schmidt farm has had to adapt to the industrial agriculture that has grown up alongside it. Long considered to have some of the most productive soil on the Atlantic coast, the Eastern Shore is also situated near some of the biggest food markets in the world.

"I'm not a 'farmers' market farmer,'" Jennie Schmidt told me. "If I took 12 million pounds of tomatoes to a farmers' market, we'd flood the market. It wouldn't work. We get better income from vegetables than from grain production. But we have the markets for diverse crops. Where in Iowa are they going to sell cannery-grade tomatoes? A hundred miles from a farm in Iowa, you're still in the middle of nowhere. Here, in less than a hundred miles, we can be in Washington, D.C., Baltimore, or Philadelphia. We can be in New York City in three hours. Without the cannery and trucking infrastructure, we wouldn't even be in vegetables."

The Schmidts' farm, while not enormous, still sits at the very center of the industrial food system, and not just because of their tomatoes and the trucking infrastructure. Their corn and soybeans go straight into the maw of giant agribusinesses, overseen and orchestrated by some of the largest and most influential companies in the world. The Schmidts' GM seeds are engineered by companies like DuPont Pioneer, and they use chemical herbicides (like glyphosate) first designed by companies like Monsanto. And their harvested beans are sold to companies like Perdue, which, in addition to being one of the world's largest chicken producers, is also one of the country's largest grain companies. It has to be: the company has to feed most of the 569 million chickens that grow on the Eastern Shore alone.

But the Schmidts also win awards for environmental stewardship. They practice a wide range of soil conservation techniques that would please even the crankiest environmentalist. Rather than strip their fields bare after a harvest, they leave withered plants to serve as "green manure." They rotate crops. They use integrated pest management. Because they spray their weeds, the Schmidts don't have to till their soil, which means they can reduce their carbon footprint—both by driving their tractor less and by leaving carbon in the soil, where it belongs. They plant cover crops, which both hold their soil in place for future crops and prevent erosion. This prevents both soil and the nutrient phosphorous that attaches to it from running off into the Chesapeake Bay. To avoid StarLink-style contamination of their soybeans, they flush out their combines and grain elevators whenever they harvest to make sure none of the GM beans intended for the chicken feed market get into the seeds grown for the tofu market.

"The truth is, you will never get to zero," Jennie said. "You can't get rid of every soybean in your combine. But there are those of us who take the time to meet that level of due diligence. Our tofu beans don't get labeled by the Non-GMO Project, but they get tested, and they are non-GMO. If they find half of 1 percent, they get sold on the Perdue market. If we can't verify that they are non-GMO, they don't get sold that way. Most people don't think we pay that level of attention. They think we're just blowing smoke, which is sad."

Given this level of scrupulous attention to soil health, conservation, and best weed- and pest-control practices, the Schmidts' farm has been certified by the state for its "agricultural stewardship," meaning it has met high standards for preserving soil and water quality.

The Schmidt farm—industrial but local, pro-GMO but pro-sustainability—offers a glimpse at a kind of middle way farming that employs technology at a scale that minimizes many of the ills associated with corporate agriculture. Their approach doesn't answer all questions, like whether we really need to be raising 569 million chickens on some of Maryland's best farmland, or whether we need to be eating so many chicken nuggets in the first place. But until we tackle those larger questions, farms like the Schmidts' suggest that the secret to better food production may lie not with enlightened global agribusinesses, but with enlightened local farmers.

Providing the Crops That Industry Demands

Jennie Schmidt received her bachelor's degree in nutrition and food science, with a minor in international agriculture. She spent a couple of years with 4-H, teaching agricultural techniques to schools in Botswana. She later returned to graduate school for a master's degree in human nutrition; her thesis looked at food and biotechnology just as the GMO industry was finding its legs. Today, she is the only woman on the board of the Maryland Grain Producers Utilization Board, and now the first female president of the U.S. Wheat Foods Council. She is also past president of the Maryland Grape Growers Association and past chair of the Maryland Farm Bureau's Specialty Crop Committee.

Now that her husband, Hans, has been appointed Maryland's assistant secretary of agriculture, Jennie relies more than ever on help from her in-laws and her brother-in-law. Still, Jennie herself has to do a lot more than just manage the family's twenty-two-acre vineyard, oversee the farm's crew, and keep the books. She also maintains a blog, *The Foodie Farmer*, on which she spends a lot of time trying to disabuse people of their fears over GMOs. Given the intensity (and in Jennie's opinion, the ignorance) of opinions on the topic, Jennie got into the GMO debate reluctantly. "I just started writing about biotech this last year," Jennie told me. "I didn't want to bring that into my home. When you criticize a farmer for what they do or don't do, you're criticizing their home. In the blogosphere, it gets very personal."

The Schmidts' neighbors, who run a 350-acre farm across the way, sell their vegetables to local supermarket chains like Giant and Whole Foods. They run a farm stand. When people think about "local farmers," it is the Schmidts' neighbors they have in their heads, not the Schmidts. This is a source of constant irritation.

"On Facebook, whenever a friend says, 'Support your local farmer,' I always chime in and say, 'You know, if you buy canned tomato products, I was one of the significant growers,'" Schmidt said. "That's the disconnect—our faces are not on those products. People don't know who we are."

This "disconnect" between consumers and farmers lies at the very root of the GMO debate, Jennie Schmidt says. Forget about gene sequencing—plenty of people don't even understand that potatoes and carrots come out of the ground.

Jennie's father-in-law, who still lives across the road, started farming in the 1930s, about ten years after the introduction of hybrid corn, which dramatically boosted the yields growers could get from their fields. Eighty years later, GMOs offer Schmidt Farms a similar boost. Using Bt corn, the Schmidts now get 221 bushels of corn per acre—more than 35 bushels (and $100) per acre more than they get for non-GM corn. It would be hard to persuade a farmer to give up such advantages, and indeed, it is this margin—more corn grown on the same acreage—that has made farmers enthusiastic about GMOs since they were first introduced thirty years ago.

In recent years, the Schmidts have started planting a new GM soybean engineered by DuPont Pioneer called the Plenish. In addition to being good chicken feed, the Plenish beans, when processed for their oil, create a second market, for fast-food frying oil. Oil made from Plenish soybeans has zero grams of trans fat and 20 percent less saturated fat than hydrogenated vegetable oil, and is high in oleic acid. The oil is also "shelf stable," and so is especially useful in the creation of processed foods that sometimes sit on store shelves for weeks or months. Perdue can use the soybeans to feed its chickens and then process the soybean oil to sell to the fast-food industry, which sees GM soybean oil as the future of fried food.

"High oleic soy can help reduce lots of health problems, because if you don't have high oleic oil, what you need to do is hydrogenate the oil to be suitable for frying and other cooking, and when you hydrogenate oils you end up with something that's conducive to cardiovascular disease," Paul Anderson from the Danforth Center told me. "High oleic doesn't have the instability that requires hydrogenation. It's very beneficial."

The Plenish beans have clearly been a boon for the Schmidts: they clear $263 per acre for these beans, compared with $124 for feed beans and just $62 for the beans they grow for direct human consumption, like the tofu market.

In other words, GM seeds are good for the Schmidts. And because the grains they grow help form a significant block in the foundation of the country's industrial food system, from its chicken nuggets to its french fries, the GM seeds are also good for the many food industries that use them. With GM products like the Plenish bean, fast food and processed food will be a bit less unhealthy. To those overseeing this industrial food system, this is a good thing.

"We've had folks ask us, 'Why didn't the industry get started with a biotech product like this?'" Russ Sanders, the director of food and industry markets for Pioneer, has said. "We think it's a great opportunity to help illustrate the positive aspects of biotech that go beyond farmer benefits."

Clearly, the companies that both provide the seeds and buy the beans from the Schmidts think the system is working. In the fall of 2014, DuPont Pioneer and Perdue AgriBusiness announced that Perdue would more than double—to about 50,000—the acreage contracted to Eastern Shore and Pennsylvania farmers for growing Plenish beans "with the intention of marketing the high oleic soybean oil by the food industry in 2015." Nationwide, the United

Soybean Board has set a goal of 18 million acres of high-oleic soy-beans by 2023, which would make the beans the fourth largest crop in the United States, behind corn, conventional soybeans, and wheat.

The move to expand Plenish beans in the Mid-Atlantic was hailed as "an important milestone for Pioneer in its efforts to bring product innovation to the food industry and complements solutions offered by DuPont Nutrition & Health to address the world's chal-lenges in food."

"We're always looking for ways to bring new market opportuni-ties to our grower customers," a Perdue AgriBusiness vice president said. "By working with DuPont Pioneer on the production of Plen-ish high oleic soybeans on the Eastern Shore, we're hoping to generate additional profit opportunities and long-term industry growth."

So here we are again: GMOs have always been pitched as "good" for farmers, and for farmers like the Schmidts, this is plainly true. They are also clearly good for the companies that make them. But are the foods these grains produce good for the rest of us? Processed foods fried in high-oleic-acid soybeans, after all, are still processed fried foods. Beyond this, are tens (or hundreds) of thousands of acres planted with these new seeds good for the environment?

Jennie Schmidt can't control the first question, but she can con-trol the second. She has no interest in telling people what they should or should not eat. If the market demands Roma tomatoes, she will grow them. If the market demands high-oleic-acid soy-beans, she will grow them. And she will do it in as sustainable a way as she can. And to her mind, GMOs help with this.

Jennie and Hans first started using GM seeds in 1998, and like many GM farmers, they maintain that the crops—which are de-signed to withstand the herbicide glyphosate—have allowed them to dramatically reduce their use of harsher pesticides, like atrazine. "We've been farming with GMOs for seventeen years and have seen a real benefit," Jennie told me. "A real reduction in the volume of pesticides for Roundup Ready crops. Using Bt corn has also eliminated a lot of insecticide use. We're using softer chemicals and using less of them."

The Schmidts consider their farm synergistic, in that it uses techniques from all three forms of agriculture: organic, conventional, and biotech. Because they plant GM crops and use synthetic pesticides, the Schmidts farm cannot be certified as "organic." The Schmidts still use some atrazine on their corn, to suppress weeds, and—given the forty-five inches of rain that falls during the growing season—they have to spray their vegetables with fungicides "just to deal with mold," Jennie said.

"There is no 'one' system that is 'best,'" Jennie wrote. "There is no 'one' way of doing things that should be done carte blanche by every farmer, everywhere. There is no 'cookie-cutter' system that should be applied to every farm. What we farmers should be doing is maximizing the synergies of all best management practices that meld together the best for our soils while preserving our inputs and natural resources."

Jennie Schmidt speaks with honesty and precision about all parts of the growing process on *The Foodie Farmer*. She explains the difference between spraying and dousing, noting (right down to the ounces per acre) exactly what kind and how much fungicide she and Hans apply to their fields. In one post, she showed her readers a photograph of a paper towel she laid down alongside a row of grapevines just before she drove by with her tractor-mounted sprayer. After passing over the towel with the sprayer, she snapped another photo. The towel is speckled, but far from drenched.

"Because this is spraying and not dousing, I do not need to soak the paper towel," she wrote. "The plants do not get 'doused.' There is no dripping off of chemical solution. They do not need to be soaked in herbicide to achieve good weed control. There is no saturation. There is no dousing."

This is precisely the kind of transparency that activists tried (in vain) to wring from the big companies on Kauai and Maui. The gamble for farmers like the Schmidts is that consumers will be willing to buy produce grown with chemicals—or with GMOs—as long as they trust that their farmer is both skilled and forthcoming about the work that goes into growing their food. For the Schmidts, this approach clearly seems to be working.

Besides, Jennie told me, it's not like the "natural" pesticides used by organic farmers are benign. The Schmidts have had as many as 100 acres of certified organic fields in the past, and even then (and even today) they still used "organic" fungicides like sulfur and copper sulfate that are, in the strictest sense, toxic.

"The copper and sulfur we use on our grapes are 'natural,' but they are still very toxic," Jennie told me, noting that nicotine, which tobacco plants generate to protect themselves from insects, is also a "natural" pesticide. "If 'natural' were safe, then smoking would be good for us," she said. "I would love a GMO grape. Sulfur is not a fun product to work with."

Especially for a farmer who has experience working organically, Schmidt has very little patience for the way some companies market the "organic movement." She now considers the label "organic" to be little more than a cynical marketing ploy that ends up making food more expensive than it needs to be and—worse—pits one kind of farmer against another.

Take Chipotle. In 2015, the fast-food burrito franchise received both praise and rebukes for announcing that it would provide GMO-free ingredients in its many restaurants. This was initially hailed as a victory by anti-GMO forces, until they learned that Chipotle would continue to sell both soda (made with GM corn syrup) and meat (raised on GM corn and soybeans). Pro-GM farmers and scientists were equally appalled, but for far different reasons.

Chipotle's campaign creates "a disservice to American farmers," the Danforth Center's Jim Carrington told me. "It creates the impression that there's evil farming and happy idyllic farming, and they source their meat from happy farms. That's simply marketing. Science has shown that the feelings they are marketing are not grounded in reality.

"In general, people who have not come from a farm have notions of farming and agriculture that are romantic," Carrington said. "The wholesome farm with happy cows and all that. But the organic industry is a very advanced and well-organized industry that has grown in part by having a villain, and the villain is conventional agriculture. Organic claims to be much better than conventional and commands a price premium. It's a very lucrative premium and is in part defended by marketing campaigns, blogging campaigns, websites, and many other ways with the intent of seeing the organic industry increase in size. There is money to be made."

Indeed, like virtually every farmer I spoke with—organic, conventional, or GMO—Jennie Schmidt practically spat when I asked her about the marketing campaigns that tout a food company's "values." She laughed bitterly when I asked her about the posters in Whole Foods that inevitably portray farmers as beautifully tanned models with a bunch of carrots in one hand and a smiling baby in the other.

"When Whole Foods or Chipotle runs these ad campaigns saying they only use food from 'farmers with values,' it's just like, 'Really? You have to throw everybody under the bus to further your own marketing campaign?'" Schmidt told me. "Painting everyone as bad except for the people they do business with? That's really frustrating. That's led a number of us to become more vocal and more transparent. We have to say, 'That's not true for me.'"

"If All Farmers Used GMOs Like I Do, We Wouldn't Have These Problems"

Like Jennie Schmidt, Steve Groff is a third-generation farmer; he works his acres ninety miles to the west, in Pennsylvania's Amish country. Groff's grandfather started growing tomatoes in the 1950s; his father grew pumpkins in the 1970s. Like Jennie Schmidt, Groff has won awards for his environmental stewardship.

And like Jennie Schmidt, Groff grows everything from non-GM sweet corn and tomatoes to (in his case) 165 acres of GM corn and soybeans.

Groff plants Liberty Link soybeans and uses a sprayer mounted on a tractor to spray glyphosate to kill dandelions, Canadian thistle, hemp dogbane (a heavily rooted perennial weed), and annual rye grass. Glyphosate "does a really good job," Groff says.

Groff doesn't grow herbicide-resistant corn, but he has planted Bt corn, which stands up better because the corn borers can't get it. That's where you get the argument that GMOs help farmers use less pesticides, he said.

"The success of Bt corn—they've really knocked down the corn borers nationwide," Groff told me. "You'd really have to have your head in the sand to dispute that. So with GMOs, I can argue why you should use them, and why you shouldn't. I tend to think objectively and scientifically. I think we need to keep monitoring this, but it's my feeling that people are overreacting to every little sniffle."

Groff, like Jennie Schmidt, thinks GMOs and synthetic pesticides have their places, provided they are used intelligently and as part of a larger, sustainable approach to farming. Groff also practices integrated pest management, which means (among other things) that he uses one-quarter of the usual amount of fungicides—and almost no insecticides.

Groff sprays some fields twice a year, and others once every two years. It depends on the rotation, he says, but probably rounds out to about once a year per field—about a third as much as most farmers, he said.

"Let's get real about this: if all farmers used glyphosate and GMOs like I do, we wouldn't have these problems," Groff told me. "They've been way, *way* overused. Sure, it's made farming easier for farmers. It's easy to kill stuff. When Roundup first came out, your crops were really clean—there were no weeds. Now you can see resistant mare's tail on this farm. It may have blown in from other farms, but so far they have not been an issue for me. I'm not sitting here worried about how to control weeds."

Steve Groff's genius lies far beyond his limited use of GMOs and pesticides. What he is really interested in is radishes—and not the kind people eat. Working with Raymond Weil, a soil and crop scientist at the University of Maryland, Groff has developed something called a tillage radish, which he considers a radically simple way to fix many of the problems created by industrial-scale agriculture.

"Farmers have nitrogen leaks," Groff told me. "If you had a leak in your barn, you'd fix it. I tell farmers they should fix the leak in their fields."

Planted as a cover crop, tillage radishes perform all kinds of jobs now done with chemicals: they control weeds, so farmers can reduce their need for herbicides. They loosen soil compaction and prevent runoff. Since the radishes pull nitrogen out of the soil and store it in their tubers, they greatly reduce the need for synthetic fertilizers: when the radishes die and break down (without the need for chemical burn down), the nitrogen goes back into the soil, where it can be taken up and used by cash crops.

"I think generally speaking we've plateaued with where we can go with chemical management styles," Groff told me. "We've seen that soils are not holding up during weather extremes. We *have* to build up soil resiliency. I'm not saying we have to eliminate fertilizers and pesticides, but we've forgotten and ignored our soil for fifty or sixty years. Some farmers are blind—they say, 'That's the way my granddad did it.' Maybe it served them well, but we're starting to see the limits of that way of thinking. That's where the younger generation of farmers can really help."

For farmers, there are no margins, Groff said; you either make it big or you go under.

"Once you understand cover crops and the value of taking care of your soil, it's like opening a savings account," he said. "By the very nature of keeping green things in the field, collecting sunlight, changing it into organic matter, it's money in the bank."

When he started planting radishes—along with nitrogen fixers like hairy vetch and legumes—Groff's soil contained just 2 percent organic matter. With cover crops now depositing up to 40 pounds of nitrogen a year into the soil, he's gotten it up to a very healthy 5 percent—all without the use of petrochemicals.

Farmers should be saving and creating nitrogen, not buying it, Groff said. "The nitrogen is here. I own it. I'm reaping the interest. And by the way, I'm not polluting the Chesapeake Bay. I don't get paid to plant cover crops, but it's an investment in your soil health. Add it up over ten years, and it will begin to pay off. In extreme weather—wet, dry, hot, or cold—having more organic matter in your soil will make your crops much more biologically resilient. I've seen thirty to forty bushels per acre increase with corn in dry season. You can have soils that are working like an IV without the chemistry."

Groff's research into cover crops is on the leading edge of a national trend. In the Chesapeake Bay watershed, state officials have pledged to nearly double the amount of farmland planted in cover crops to 460,000 acres, or roughly half of all croplands in the state.

Across the country, cover cropping has grown 30 percent over the last couple of years and is being used on everything from small organic farms to large industrial operations.

"I don't know of any other concept that's sweeping agriculture like this," Groff told me. "They're all getting into it. Eighty percent of the Amish farms around here have radishes in them. I had a guy in Illinois say he wanted to buy radishes for 1,000 acres. I said, 'Whoa—you ought to start small.' But he said he had 40,000 acres, so I said okay."

Groff's tillage radishes, which he now sells nationwide, have been so successful he now has twenty-two people working for him: selling seeds, doing marketing, conducting research, and influencing agriculture nationwide. Not bad for someone who never went to college.

"There was never a day I didn't think I was going to be a farmer," Groff told me. "People ask me where I graduated from, I tell them I haven't finished learning yet."

Like Steve Groff, Jennie Schmidt is willing to use any farming technique—GM or non-GM, pesticides or no pesticides—as long as it produces for the farm and doesn't run down the larger environment.

"We tried to tap into the organic market, but it wasn't sustainable for us," Schmidt told me. "People can't believe it when I say that. I don't mean organic is not sustainable. It's just not sustainable for *us*. There's no universal cookie-cutter method for all soil types, or all regions, or all farming sizes. Maybe if we had been less diversified, and had been just in grains, we could have focused on organic farming.

"If we didn't have so many irons in the fire, and weren't trying to sustain a family farm, and had more time to focus on things like doing organic practices, maybe that would have worked for us. It's a balancing act: this is what works for our farm and our level of risk that we were willing to take. Soil types dictate a lot of it. Smaller organic operations can overcome that, because they plant a smaller variety of many different things and have smaller acres to take care of. When you have 200 acres and a variety of different crops going, you need a very different approach."

Organic? Sustainable? Or Regenerative?

These ideas sit very nicely with Brian Snyder, head of the Pennsylvania Association for Sustainable Agriculture and one of the country's leading voices for farming done with the health of the planet in mind. Snyder wonders whether produce grown by farmers like Jennie Schmidt and Steve Groff might be given its own label: not "GMO" or "organic," but "sustainable." Marking food as "sustainable" would reward farmers for preserving topsoil, for example, or building soil quality through the use of nitrogen fixers, cover crops, or composting—even if they used GM seeds.

"People accept that 'organic' is never going to include GMO, but the question is whether the term 'sustainable' can include GMO," Snyder said. "The problem is, none of these things are simple. Compared to 'organic,' 'sustainable' offers a bigger tent that includes more than organic produce. No-till agriculture would qualify, and no-till farmers mostly use GMO seeds. We might have to consider a world with GMOs, to save soil and build soil quality. Instead of labeling foods with a skull and crossbones, they'd rather have a 'sustainable' label that is a positive thing, even though it may have GMOs in it. That idea is not completely without merit."

GMOs themselves will never cause a fraction of the problems caused by the industrial food system itself. "We are going to waste a lot of time and energy on whether GMOs are helpful or harmful to people, when that's not the most important question," Snyder said. "The most important question is what kind of *system* do they generate and support? The most effective criticisms of GMOs are about the peripheral realities of this system. It used to be that farmers always retained a percentage of their crop for seeds for the next year. They did this for thousands of years. Now, in the last couple of decades, almost none do that."

These ideas also sit nicely with Blake Meyers, the University of Delaware geneticist, who thinks GMOs may one day be a key component in sustainable agriculture. "I buy and eat organic food, and I don't like chemical residue on my food," Meyers told me. "So the question is, why should you not have organically grown GM crops? So far the USDA has said you can't have GMOs and label them organic, but I would prefer to get rid of all the chemistry and confer the desirable traits with genetics. Petroleum won't be with us forever. I'm perfectly okay with transgenics: imagine if we could create a wonder crop that requires genetic modification but is grown organically, free of chemical inputs, and resists drought, resists pests, and outcompetes weeds. I'd be okay with that."

Some of the critics of GMOs "are the same people who spray copper or sulfur on their plants and say it's okay because it's 'organic,'" Meyers said. "They apply Bt to their crops, but God forbid you take the same genes they're eating in the bacteria and insert them into the plant. Ultimately, our agriculture has got to be sustainable, or we're not going to be here long term. GM will be a large part of that. Fifty years from now, we'll see that spraying anything on plants created a huge amount of waste and pollution. We'll see that 90 percent of the chemicals washed off, ending up in our soil and water. If you eliminate that waste, if you can use GMOs to replace these inputs, achieving similar yields, without all the chemical inputs, you've done a world of good."

There are legitimate worries, of course, that a comparatively loose term like "sustainable" could be distorted and abused—"greenwashed"—by industrial farms in ways that a strict, legally precise term like "organic" cannot. The word "sustainable" has been "overused, misused, and it has been shamelessly co-opted by corporations for the purpose of greenwashing," write veteran food activists André Leu and Ronnie Cummins. Indeed, they note, the word is featured prominently on Monsanto's website, where the company boasts of a "commitment to sustainable agriculture—pledging to produce more, conserve more, and improve farmers' lives by 2030."

"Industrial agriculture today, with its factory farms, waste lagoons, antibiotics and growth hormones, GMOs, toxic pesticides and prolific use of synthetic fertilizers, doesn't come close to 'not using up or destroying natural resources,'" Leu and Cummins write. Instead of "sustainable," they would like to see foods affixed with one of two labels: "degenerative" or "regenerative." Consumers could then choose food produced by chemical-intensive, monoculture-based industrial systems that "destabilize the climate, and degrade soil, water, biodiversity, health and local economies," Leu and Cummins write. Or they could choose food produced using organic regenerative practices that rejuvenate the soil, grasslands, and forests; replenish water; promote food sovereignty; and restore public health and prosperity—"all while cooling the planet by drawing down billions of tons of excess carbon from the atmosphere and storing it in the soil where it belongs."

Such rhetoric—powerful, convincing, and justifiable as it may be—is plainly directed at giant corporate farms, and the global food companies they serve. The question for midsize farmers like Jennie Schmidt and Steve Groff is: Can they operate inside this system in a way that is more benign? And can GMOs be a part of this? As with everything else in the American food system, there are no simple solutions: as enlightened as Jennie Schmidt and Steve Groff may be, they are still plugged into a larger food system that uses enormous amounts of chemicals on vast swaths of land to create huge quantities of unhealthy food. But it's a start.

To Jennie Schmidt, this global food system—with all its downsides—will continue to evolve because people will continue to enjoy foods they can't grow themselves.

"I'm not going to give up coffee or chocolate," she told me. "I love the fact that we have so many food choices. Yes, there are downsides, but there are lots of upsides to it too. I like the fact that in February I can go to Millington, Maryland—two miles away is the closest grocery store—and get fresh produce. Think about what we'd have to do to grow that produce around here, in winter, in hoop houses, and at what cost? When you have to use so much propane to heat the hoop house, that can be more energy-intensive than getting it from Mexico.

"My concern with people's resistance to the technology of GMOs is that the next generation of products, and the next round of benefits for folks in developing countries—for traits they need to resist certain diseases or yield—don't come about because they are not *allowed* to, because there has been so much pushback," Schmidt said. "All of plant breeding, whether it's traditional or GMO, has benefits, and my concern is that science will be stifled because there has been so much resistance to it. Think of what we could do if we could get rid of food allergies from soy or peanuts. If you could silence the protein that causes peanut allergies, that would be a big deal. I'm afraid we're going to throw the baby out with the bathwater."

The Farm Next Door

There are two spiritual dangers in not owning a farm," Aldo Leopold wrote in his landmark book *A Sand County Almanac*, first published in 1949. "One is the danger of supposing that breakfast comes from the grocery, and the other that heat comes from the furnace."

Almost seventy years later, these dangers—as Jennie Schmidt knows all too well—are clear and present. As suburban sprawl has continued to eat millions of acres of the nation's prime farmland, fewer and fewer of us—especially those of us living in and around large cities—ever have a chance to connect with the way our food is grown.

When it comes to our food, we are all blind, even if it is for different reasons. If we live in the city, we rarely have the chance to see where (or how) food is grown. Ditto for the suburbs: even if there were once crops occupying the fields where subdivisions now sit, they aren't there anymore. Even in rural America, where there are plenty of farms, it's hard to get your eyes on actual food: the corn and soybeans growing across America are sold to industrial food processors, feedlots, and energy companies, not on farm stands or in supermarkets.

But it's important to remember that industrial-scale farming (with or without GMOs) has been around for only a few short decades; before that, everyone ate what they—or their local farmers—grew for them. And today, in small places all over the country, this sort of farming—small, local, often organic, and decidedly *not* GMO—is once again beginning to sprout. On small farms in the country, in the suburbs, even in run-down sections of industrial cities, small farmers—responding to a growing unease with industrial food production—are beginning to connect with the people they feed. In the process, consumers are not only paying more attention to their food, but paying more attention to their farmers and to their land.

Outside the City

Drew Norman is entirely sick of hearing that GMOs are the future of food. Sure, he knows that by the end of the century, the world may well have to accommodate an additional 2 billion people, many of them in the developing world. He knows we will need to grow more food, on less land, in the most efficient way possible. So, are genetically engineered foods the answer?

"Fuck that," Norman told me.

A lanky, graying man in a seed cap and leather boots, Norman is no dewy-eyed environmentalist. He is an avid deer hunter, with trophy heads mounted on his wall. He has few positive things to say about government regulators and considers local environmental groups to be obnoxious to farmers. His son is scaling up the family hog farming business from eight animals to several hundred.

But Norman is also the owner of One Straw Farm, one of the largest organic vegetable farms in Maryland. Norman started farming thirty years ago and has been running his farm as a community-supported agriculture (CSA) operation since 1998. Working a 65-acre vegetable farm, plus another 150 acres of forest and hayfields, he and his wife, Joan, now supply food to some 10,000 people a week, through their 1,900 CSA members and by delivering to a half-dozen Maryland farmers' markets. In his barn, the day I visited, there were stacks of crates filled with squash, tomatoes, and kale. Four large trucks were preparing to deliver to thirty-eight pickup CSA sites around the state, including some right in the heart of Baltimore.

Margins are small for any farmer, and Norman has had to be nimble. His team has been selling canned tomatoes and peppers for years; they are now doing Bloody Mary mix and are about to market a tomatillo salsa. His son is planning to raise two hundred hogs that will feed on acorns in the farm's oak forest. Chickens may be next.

Norman's counteroffer to corporate, monoculture, GM farming is simple: Buy local. When you buy a cabbage from your local (and preferably organic) farmer, you don't have to worry about whether it's been "tested," because the food was grown the way food has always been grown (or *had* always been grown, before petrochemicals and genetic technology entered the equation): with seed, soil, sun, and rain. You also don't have to worry about whether part of the cost of the cabbage is going to pay the salaries of seed company executives in an office building in St. Louis, or a political lobbyist in Washington, or a pesticide company in Wilmington. Your money goes to the farmer.

If there were more farms like his—and until just a few decades ago, there were—none of us would need to eat engineered food,

Norman says. His food, organic and local as it is, is also cheap. I pay him $500 every winter for six months of produce in the spring, summer, and fall—which works out to about $20 a week for a large canvas bag stuffed with everything from lettuce, spinach, and collards to acorn squash, sweet potatoes, and watermelon. This is routinely more than my family of four can eat.

Norman runs a farm that is a model of sustainability. He intercrops (strawberries with oats, for example) to prevent erosion. He plants cover crops. He has enormous windrows of compost. His feelings about raising meat mirror his feelings about farming generally.

"There was an equal number of buffalo when we got here as there are cattle today," Norman told me. "They were eating and doing their thing and providing meat in a pretty environmentally friendly way. If we planted our corn into grass and raised cattle on it, we'd probably have the most environmentally friendly way to produce protein there is."

In Europe, where small farms like Norman's have been the model for hundreds of years, opposition to GM crops has been intense since the beginning. Although typically framed as an issue of food safety, Europe's anti-GMO argument is also fundamentally built on anxiety—or outright anger—over the effect of large-scale farming on small-scale farmers. Italy, France, Spain—they have all spent centuries stitching together small-scale farm economies and take well-earned national pride in the quality and integrity of the food these farms produce.

But given that by the end of the century less than 10 percent of the world's population will be living in Europe, is a European-style, small-scale agricultural model something the rest of the world can afford to emulate?

Drew Norman, and many others, think the answer is unequivocally yes. Not only would the food this system creates be healthier,
it would support local economies and curb the power of global food
conglomerates. Regardless of whether GM foods are "dangerous,"
they are definitely corporate, aggressive, and—in every sense of the
word—monopolizing. They so dominate the agricultural, political,
and cultural landscape that consumers—here and abroad—can't opt
out even if they want to.

To Norman, the heart of the issue is trust. "If something is being
tested by the people making money from it, I don't trust the tests,"
he said. "If there's no government oversight or independent testing
on the safety of a product, I don't trust the company—who's going
to make millions or billions—to be honest."

But can farmers like Drew Norman really feed all of us? Especially given our current eating habits—more bags of chicken nuggets than bundles of organic kale—does the world even *want* what
Drew Norman is growing? Local, organic food, to many people,
seems like a yuppie indulgence: boutique, expensive, and—in the
end—a lot less satisfying than a burger, some fries, and a Coke.
And if this is true in the wealthy United States, isn't it also true in
the developing world, where companies are hard at work pushing
their GM corn and soy?

"Look," Norman said. "The Third World can't afford to buy
Stouffer's meals. The Third World needs to use local ingredients
and cook in their own kitchens. A local food system is a really easy
thing to do in the Third World. That's what they've always done,
and that's what they're doing right now. Kenya is number three in
the world in certified organic farms. They are pretty food sufficient,
and they've done that by supporting local food systems and local
agriculture."

As for American consumers, who claim they "have no time" to cook their own food—let alone think about how (or where) it is grown: "Quite frankly, by the time you drive to McDonald's and buy your burger and fries, you could have made your meal at home," Norman told me. "Americans are so busy chasing an income to be two-percenters or whatever it is, they don't have time to look around at the environment or their health. All they have to do is look in the mirror, but they're so busy doing what they're doing they don't have time to do even this."

I asked Norman about national trends in obesity, diabetes, and all the other ills associated with GMO-driven fast food and processed food, and the next-gen GM products that promise to deliver these same foods with less fat, salt, and sugar. "It all goes back to money," Norman said. "The likely solution is the stupid solution. GMO potato chips? That's not the solution. The solution is to eat more fruits and veggies, not the thousand-calorie coffee drink. That's where I just think Americans don't look at what they're doing."

We have to question a food system that puts way more energy into food production than we're getting out of it in food calories, Norman said. Conventional farms can use 10 calories of energy to produce 1 calorie of food—and this is *before* calculating the energy it costs to ship food around the country and the world. Food grown in the South and Midwest travels an average of 1,500 miles from farm to plate.

With the exception of some feather meal he buys from nearby chicken farmers, Norman grows all his own fertilizer through the use of compost and nitrogen-fixing cover crops. A head of lettuce grown with petrochemicals and shipped from an industrial farm in California to Baltimore, in other words, is a far more polluting (and far less efficient) vegetable than lettuce grown with no chemical inputs and shipped twenty miles from Norman's farm.

"Scientists reduce everything to its simplest forms, but they are not looking at the big picture," Norman told me. "You can do a lot in the laboratory, but it needs to be rounded out by people looking at the big picture, and talking to each other. If the world looked at the problems associated with industrial agriculture as a whole, they would realize the food is not inexpensive. We need ethicists and ecologists to be a part of this conversation. We have to stop looking at everything as only coming down to the bottom line."

In the Country

Three hundred miles to the north, among the rolling hills and dairy farms of New York's Hudson River Valley, Steffen Schneider is doing everything he can to return farming to its rightful place as the center of community life. Schneider, like Drew Norman, thinks that GMOs are merely a symptom of the invisibility of food production.

"I always come to the conclusion that GMOs are an answer to the wrong question," Schneider told me. "They always say, 'We have to feed the whole world,' but that's not the right way to approach it. Clearly, there is already enough food to feed more people than are alive right now, so that's not the right question. In my mind you would have to look back and reflect on agriculture's role as *humans in relation to food and to the planet.* When you have a compass and you're trying to figure out the way to go, you don't find out by taking the compass apart and breaking it down to its atomic structure. That's not going to get it."

Schneider works 400 acres at Hawthorne Valley Farm, ten miles west of the Hudson River about two hours north of Manhattan. His farm includes 15 acres for vegetables and 40 acres for grain. He grazes sixty dairy cows and thirty beef cattle, and he keeps as many as forty hogs. Schneider runs Hawthorne Valley as a "biodynamic" operation, following principles put forward by the Austrian visionary Rudolf Steiner. Long before the modern organic farming movement, biodynamic farmers paid scrupulous attention to the intertwined ecological health of their entire agricultural system, from soil and weather to plants and livestock. Synthetic chemicals are anathema, as is the kind of mistreatment of animals that has become such a grim trademark of industrial-scale feed and slaughter operations.

"There needs to be an inner shift—that's been my recognition these last few years," Schneider told me. "People are looking and searching. A big part of our customer base here is mothers and young people worried about feeding their families. Hopefully, over time, we will make these changes. Otherwise, we're just going to stay stuck. What are our responsibilities? Everybody has to ask this of themselves. There is an amazing opportunity to ground this change in agriculture."

Schneider got a degree in agronomy in Germany before coming to the United States in 1983 (he worked a dairy farm in Wisconsin for seven years before moving to Hawthorne Valley). Following Steiner's biodynamic ideas, he imports nothing to the farm except tractor fuel and electricity: his fallow fields are protected and enhanced with cover crops like rye, vetch, and red clover; he grows his own hay and produces all his own fertilizer from compost and animal manure. From April until late fall, his cows are in pasture, and his pigs are fed whey during the cheese-making season, food scraps from the Hawthorne Valley grocery store and deli, leftovers from his own sauerkraut, and milk by-products. The closest meat processor is just twenty miles away.

For Schneider, a farm should be the centerpiece of a local economy, not a cog in the global economic machine. His fields are part of Hawthorne Valley's larger vision for "social renewal" that includes—right next door, in the same beatific valley—not only a sizable farm store but also a K–12 Waldorf school, a Place-Based Learning Center, and a Farmscape Ecology Program. Schneider's farm provides work for eighty people on the farm itself, and two hundred if you include the store and the school. "It's a great thing to be able to offer meaningful work to so many people," Schneider said.

Farmers and teachers work closely with families throughout the region to connect them with their food as well as their bioregion. Living and eating near Hawthorne Valley, it would be impossible *not* to know where your food came from, who grew it, or under what conditions. It would also be impossible not to understand the relationship of the farm to the larger landscape.

"However you define 'sacred,' essentially it's a place you love," Conrad Vispo told me. Vispo, a PhD wildlife ecologist, and his wife, Claudia, a PhD botanist, run Hawthorne Valley's Farmscape Ecology Program, an education center committed to documenting the human and natural history of the region's farms and wild landscapes. Teaching farmers, and children, and everyone else about the ways food production fits into a larger ecological context is a sure way to bring people closer to their food—and to open their hearts to the places they live, Vispo told me.

"Why do you love a place? Your experiences as a people, your individual experiences—you make your judgments based on facts, but also on your core feelings on what is right and wrong," Vispo said. "What we hope to do with our program here is make the land sacred to more people, in the sense of getting more people to love the land. Then the way they think of the land will include more than just how to use the land. It will include how the land will be affected by their actions."

In other words, Hawthorne Valley is as far from an abstracted monoculture industrial farm as it can be. The food, and the farm, and the people—they are all intimately connected. And intimacy, Steffen Schneider said, is the best way to ensure that both people and land will be properly cared for.

"In any country, the first thing they have to ask is, 'What do we have to do to develop farming systems that are successful *right here*?'" Schneider said. "The Green Revolution did a lot of good, but it was also extremely destructive because it destroyed a lot of traditional farming systems. Most people in the world are still eating local food, but these companies are saying this traditional way of farming and eating is not 'modern.' This is not helpful."

No matter where a farmer is working the land—in New York, in Maryland, or in Kenya—"we need to figure out what local adaption means right there," Schneider said. "We are one human community. We have to look at all of us as one human community. There will still be crops that we share—coffee and chocolate won't happen unless we bring them in. But we need to feed our own communities. We have to envision things radically different. We can't just do it slightly better. Then maybe this whole GMO discussion might just go away."

Hawthorne Valley functions as a nonprofit, but its business model is still highly sensitive (and responsive) to market demands. The farm produced New York State's first organic yogurt, and now that educated foodies have gone in big for fermented foods, Schneider is producing eight different varieties of sauerkraut. The day I visited, Schneider took me down to the kraut cellar, where workers were producing kimchi from Napa cabbage using a "vicious" home-made hot sauce.

The market to which Schneider has to respond ranges from people in his own rural community to five different farmers' markets in Manhattan. He sells raw milk to locals (state law prevents him from selling it off-site) and sends trucks two hours to Manhattan every Thursday to feed 250 families who are members of his CSA. His organic yogurt makes it all the way to markets in Maryland. In the winter, his greenhouse—heated by radiant hot water piped beneath the soil—produces salad mix and microgreens, both of which collect high prices in the Manhattan markets. "It's amazing what this stuff commands in NYC," Schneider said. "It's not like any of us are getting rich off this, so I don't feel bad about it."

Schneider is thinking about expanding his operation to a "full-plate" CSA, with bread and cheese and meat. For years, the farm did not raise chickens because the birds require "inputs" of feed, but now that Schneider is growing grain, he may add them. Again, it's all about balance.

"There would be *lines* in New York City for our eggs," he said. "People also want us to grow more veggies. We could, but then we would need more animals, and it might throw the whole balance out of whack."

Handling the pressure to grow is a mixed blessing for Schneider: if it's forcing him to recalculate the balance more production would require, it's also confirmation that his style of farming is catching hold. He is constantly being asked for advice by young farmers, who see his integrated, even philosophical approach to farming as a far more exciting prospect than growing endless acres of GM soybeans. To the new generation of farmers, Schneider's approach offers more than a job—it engages their imagination.

"With industrial agriculture, people practicing farming are looked at as having no social standing," Schneider said. "Farmers have become cogs in this industrial system. They aren't happy about it, but they don't feel that they have a choice. We've forced them to produce as much corn and soybeans as cheaply as possible, and they just do it, because they're stuck. Entire communities have been wiped out in the service of industrial monoculture. What's going to happen in twenty years? Agriculture can offer so much by reinventing economic principles, and our relationship with the natural world, and with each other. It's a very exciting time for me to be in agriculture right now."

Schneider's fierce commitment to local food production goes "beyond organic," and mirrors Alika Atay's argument on Maui, where schoolchildren are still fed oranges grown in Florida rather than papaya grown next door. It also mirrors Drew Norman's argument in Maryland, where local producers still struggle to get their produce into conventional supermarkets. Their common push is to find ways to support food that is grown—and eaten—within individual communities.

The market for local food is plainly growing. Nationwide, the number of farmers' markets has increased 76 percent since 2008. In Delaware, where I work, farmers' markets have increased more than eightfold since the state Department of Agriculture began tracking them in 2007. Fresh produce now makes up 60 percent of local produce sales, with the remainder coming from value-added products such as meats, cheeses, jellies, breads, salsa, eggs, and honey.

"Over the last few years, we have seen an incredible rise in people wanting to eat healthy and buy fresh, local foods for themselves and their children," the state's secretary of agriculture said recently. "Our farmers and producers are working to meet that demand by selling some of the best fresh produce, meats, cheeses and honey that any state can offer. Our farmers' markets also connect the people who eat with the people who grow their food, fostering conversations and friendships that can last a lifetime."

Thanks to a national surge in demand for both organic and local food, the federal government seems finally to be getting the message. The USDA announced in 2014 that it would spend $52 million to support local and regional food systems, including not just farmers' markets but local food distribution networks, and to do more to encourage research into organic farming methods. The Obama administration has also tripled—to $291 million—federal funding for organic farming, including $125 million for research and $50 million for conservation.

As helpful as this has been, it's worth keeping these numbers in perspective. The total federal farm bill in 2014 was $956 billion, including more than $44 billion for commodity crop programs. Compared with these numbers, the money given to local or organic farm programs is barely a rounding error.

Steffen Schneider's sense is that for the local, organic sector to grow, it will have to find a way into the kitchens of people who aren't only in search of expensive microgreens. Hawthorne Valley accepts WIC stamps for their CSA. He is exploring ways to open a store in Hudson, a historic city that is popular among weekenders from Manhattan but which remains largely working class. His wife runs a program called Kids Can Cook, a three-week day camp that teaches local kids how to grow, harvest, and cook their own food.

But to think about the price of local organic outside the context of the way it is grown—and outside the context of the way "conventional" food is grown—is to miss a much larger point. In Germany, Schneider said, research has shown that as per capita spending on food goes down, health care costs go up. In other words, the cheaper your food, the worse (and more expensive) your health.

It may be true that organic farms have generally lower yields than farms that use petrochemical fertilizers and pesticides, though recent studies indicate the differences may be smaller than previously thought. Organic corn, soy, and wheat can yield up to 97 percent of crops sprayed with chemicals, one study showed; other crops, in other places, fare almost as well. Gaps can be reduced further—or eliminated completely—by growers who "mimic nature" by creating "ecologically diverse farms that harness important ecological interactions like the nitrogen-fixing benefits of intercropping or cover-cropping with legumes."

A thirty-year study conducted by the Rodale Institute found that chemical-dependent farming may outyield organic farming during good years, but over the long haul (and especially during drought years), organic systems, with their vastly healthier soils, outyield conventional systems. Organic farming also reduces the use of fossil fuel energy by about 30 percent and significantly improves the organic matter in the soil itself.

But comparing yields—pretty much the only metric that "conventional" farmers like to use—is a puny way to think about the optimal way to grow food. So is "convenience," the other word industry uses to pitch processed food.

"Agriculture is not just an economic activity designed only to produce cheap food," Schneider said. "It is a multifunctional reality that underpins all culture and economics and has ecology as its foundation. Looked at that way, farming takes on a whole different meaning."

It is political dogma to say that food has to be cheap, Schneider said; such a stance devalues both food and the farmers who grow it. A far better parameter by which to decide how food should be grown? Health—both human and environmental.

"Good food should be a right everyone has," Schneider said.

"Think how many millions of dollars people are spending to see a new movie. Some of this really requires a rejiggering of awareness. Food has been looked at as sort of an afterthought. It's only recently that people are beginning to realize that this isn't the right place to skimp. Only recently have people begun to think about the link between food and health, which is nuts. Health is really the only sensible outcome by which you can measure agriculture. And by health I mean the health of the earth, of communities, and of individuals. When you think about how effective this industrial agriculture has been, this has been a failure."

A couple hundred years ago there was a Hawthorne Valley in every community, and as late as the mid–nineteenth century, more than half of all Americans worked on a farm. Which poses the question: In our advanced technological age, and with our exploded population (thanks in no small part to expanded food production developed during the Green Revolution), is it in fact possible to live in a world without GMOs, or without industrial farming itself? Can there ever again be such a thing as community self-sufficiency, or are networks of enormous farms and global transportation systems here to stay?

In Europe, large cities like Rome and Lyon—and countless smaller cities, like Orvieto and Avignon—are ringed by hundreds of small, diverse farms. The answer in the United States lies in many more acres, in many more places, involving lots more people— and not just in the middle of the country, said Craig Holdrege, the scientist and philosopher who runs the Nature Institute, an environmental and agriculture education center just down the road from Steffen Schneider's farm.

"If you think of metropolitan New York's 16 to 20 million people, you can't have a single Hawthorne Valley feed that whole population," Holdrege said. "But the regionalization of food production—

that's really happening. You can have urban agriculture—there is a big movement in Detroit—and you can have a lot of farms around these population centers. There would be no problem feeding Chicago from right around the city if you took some of that land out of soybeans and corn. That's how it was done just a hundred years ago. The city was fed by its region."

The City Farm

Baltimore City has 11,000 city employees. Guess how many of them are farmers?

One.

His name is Greg Strella, and his 33-acre farm—situated behind a Popeyes chicken joint, a discount mattress warehouse, and a Pep Boys auto parts store—is where Strella is teaching city kids how to grow and eat things most kids wouldn't be caught dead eating. Like beets. And sorrel.

A couple of years ago, Strella, the manager of Great Kids Farm, noticed something strange as he led school kids on tours of his farm. As they walked between farm buildings, the kids kept bending over and sneaking handfuls of a perennial plant called sorrel. Sorrel grows easily along sidewalks, and Strella had planted small batches not as a crop but as a kind of edible landscaping.

"When you're standing outside with students eating lunch, and students are coming out of the lunchroom and sneaking sorrel and saying, 'Is it okay if I have more sorrel?,' it totally reverses the challenge of getting students to eat something they don't want to eat," Strella said.

"'Sure,' you say. 'You can eat more of this pure food.' It's a beautiful inversion."

Strella was not trained as a farmer, he was trained as an artist at the Maryland Institute College of Art. As a student, wherever he went—the classrooms where he studied, the restaurants where he waited tables or worked the kitchens—he was confronted by his own ignorance about food. But he kept asking questions, and every chef he talked to responded with "an endless willingness to share what they knew and already experienced."

Strella has now been farming for more than ten years, ever since he got out of art school. He went to Chicago and volunteered at City Farm, near the Cabrini Green housing projects. He tutored under Will Allen, the legendary urban farmer whose Growing Power operation has become world famous for both feeding and employing people on the farms he's built in downtown Milwaukee. This led to an apprenticeship at a small CSA farm in Lancaster County, Pennsylvania, and then a job in Baltimore, as the first farmer the city had ever hired.

The piece of land he was given did not, at first, seem promising. There were busted-up greenhouses and soil that hadn't been properly cared for in years. And it wasn't like Strella himself was bringing generations of farming wisdom to the table. He had trained as a sculptor.

But Strella was taken by other things, like the old orphans' home with historic photos of young black teenagers helping black masons put the building's stones in place. The place had a history of craftsmanship and self-determination that Strella liked a great deal.

Urban farming also drew him. Some 20 percent of the world's undernourished people live in cities, and urban "food deserts"—where food can mostly be found in gas stations or convenience stores—have become a focus of intense concern for public health experts. Millions of America's poorest city residents have little access to produce of any kind, let alone fruits and vegetables grown in their own neighborhoods. Much of this is due to the flight of middle-class residents (and the markets that served them) to the suburbs; Baltimore has lost fully 300,000 residents since 1950. Replacing supermarkets and vibrant residences, in Baltimore and Detroit and countless other cities, have been vacant lots.

But into this vacuum a new generation of inventive farmers has begun to break ground. Post-industrial cities like Detroit and Milwaukee—which are now dotted with bona fide farms, not just garden plots—have joined more prosperous places like San Francisco and Portland, Oregon, as innovative hubs for young people trying to reinvent the way food is grown, and delivered, to the country's population centers. Detroit alone has some 1,300 community gardens; Portland has twenty-six farmers' markets; gardeners in Austin, Texas, provide the city with more than 100,000 pounds of local food a year. Baltimore's mayor recently started a Vacants to Value program, intended to nurture urban renewal and promote open space; neighborhoods can lease land to create gardens and green space.

Urban farms provide local food and jobs to people who see too little of either. They also provide places of respite and natural beauty within caverns of concrete, especially for school-age kids who suddenly have recreational options beyond the pavement.

Which brings us back to the magic of sorrel. For his first three years at the farm, that became the most common question Strella got from his students: Can I please have more sorrel? So Strella did what any farmer does: he acknowledged the desires of the market and he provided. The following year, he and his students planted a thousand pounds of sorrel.

"No adult in their right mind would sit down and design a process to 'create curiosity and participation in eating salad' by serving a tart, lemon-flavored green," Strella said.

By listening to students—and figuring out how to navigate the city's public school bureaucracy—Strella had little doubt what he ought to add to the public school salad bar. "That probably seemed outrageous to anyone who hadn't just seen thousands of four- to twelve-year-olds enjoying this food," Strella said. "But to us it was the obvious decision, because we had watched our students and listened to our students."

That spring, Strella and his student farmers got sorrel into twelve inner-city lunch programs. Later, when the schools surveyed what the kids were eating, and what they would like to see at the school salad bars, 97 percent said they would eat more sorrel.

As Great Kids Farm began to take root, Strella and his shoe-string staff began organizing visits from elementary and high school students into three parts: some time in a classroom, some time exploring the landscape around the farm, and a tasting experience, where students would sample the foods they were helping to grow on the farm.

This is where Strella introduced five hundred Baltimore city kids to beets.

The idea was to find a way to get root vegetables into school cafeterias, and to prepare them in a way the kids would eat.

Potatoes, I could see. Carrots, sure. Even sweet potatoes. But beets? Aren't beets, for American kids, the universal symbol of disgust? In a bit of ironic serendipity, many of the students Strella was trying to convince didn't hate beets because they had never seen one.

"Part of the quirkiness of the urban environment is that it's almost like the preconception against beets isn't there," Strella said. "This creates different opportunities for forming relationships with food. The kids could just as easily have been out playing basketball. Here they were, harvesting beets."

Strella picked six high school seniors to plant, and tend, the entire crop. Once the beets had been harvested, Strella and his chef set up stations with the vegetables prepared five ways. ("It was kind of like a wine tasting," Strella said.) He was sure that the most promising, since it offered the most sweetness, would be beets mixed with orange juice. Station Number Five was shredded raw beets, with nothing added: no sugar, no dressing, no nothing. Strella and his chef had set the station up as a kind of control, to see just how much more kids liked the other four recipes.

To his surprise, and by a large margin, the kids picked the raw beets.

"That's just one example of what happens when we put young people in a position to be collaborators, and give them the opportunity to make decisions for themselves and to take the risk of trying new things," Strella said.

As with the sorrel experiment, the beet test turned into something magical. That season, Strella and his student volunteers planted and harvested a lot of beets—3,300 pounds of them—and sent them out to sixty schools, where students would serve them to their peers in their cafeterias. Strella's young farmers did all the marketing; they put posters up in their cafeterias—"What's a cucumber? You'll find out today!" "What's a beet? You'll find out today!"—and included a riff on lyrics from the rapper Drake: *Started from a seed, now we're here.*

Word started getting around that something special was happening in Baltimore. Strella and his student-farmers found themselves hosting a series of workshops for 146 food service directors from forty-eight states and Washington, D.C. The students toured visitors around the farm and shared some of the raw shredded beets they were growing for the city's school cafeterias.

"Here we have all these amazing food service directors from all around the country, and they are eating a raw vegetable that is simply a raw vegetable, not raw beets and sugar, or oil," Strella said. "The directors would say, 'What do we have to do? Where are the labor costs that make it work? The production cost must be so onerous, that must be why I can't see raw shredded beets showing up in our cafeterias.'

"We said, 'All we can tell you is that we have a tiny little staff of a farmer, a teacher, a chef, and an incredible group of students,'" Strella said. "We don't know why this can't exist in your cafeteria. What we do know is that our students can grow them, harvest them, put them in the cafeterias, and eat them. We can tell you that this is what we have done. We don't have access to anything you don't have access to."

For Strella, teaching city kids how to grow and eat their own produce is part of the vision shared by Drew Norman and Steffen Schneider and Wes Jackson and Alika Atay: it is teaching them the value of autonomy, of caring for their bodies and their communities and their local landscape.

"We see our students change physically," Strella said. "They develop shoulder muscles, a certain pace in how they walk, a certain confidence in how they work in teams, a certain resonance in their voice when they speak to young people, or in front of two hundred people for events. Those ripples, you can see them in the classroom and schoolyards and cafeterias. I'm talking about high school students growing food for their peers. These are seventeen- and eighteen-year-olds, and they are planting the seeds and picking the vegetables, and that is something that profoundly changes what it means when those veggies show up in the school cafeteria.

"When we hear people talking about the world the next generation is going to inherit, where are they right now? What are they eating right now? We want them to take responsibility for participating in that world's creation."

Arguing that industrial agriculture has a fixed and immutable place in our world is like arguing that because highways are the most efficient mode of transportation, all we need is highways, Strella told me. The truth is, we need sidewalks and bike lanes and side roads too, because "no one would discount the value of walking down a sidewalk and saying hi to your neighbors, and you can't do that from a highway.

"That's not to say we don't need highways. But it is to question the singular value of highways," Strella said. "We've had fifty or sixty years to test out our highways and industrial ag, and we now have the vista to see things we couldn't see thirty years ago. So often the debate is an incredible narrowing to cost-benefit analysis. What we're taking about is holding ourselves in relation to a much fuller accounting of the role we play in the world. In the food landscape, even just to narrow it down to 'food' is already too narrow an accounting. In our homes, in our backyards, on our streets—there is incredible value to having our food living with us, even before it gets to our plate. Once you start going up to levels of economy, you start to lose this."

Seen from this vantage point—teaching people to think as intimately as they can about the relationship between their bodies, their food, and their soil—the prospect of giant agribusiness seems entirely counterintuitive, Strella said. There are far too many externalized costs—from monoculture and pollution on the farms themselves to obesity and diabetes in the inner-city people who eat them—for GMOs and industrial food generally to make any sense.

"It's absurd that we talk about industrial agriculture as 'conventional,'" Strella said. "As a tool within the industrial system, biotech is propping up systems that are already unraveling. Think of the dead zones in the water, the loss of topsoil, the health of our bodies—biotech is right there at the center of all that. Is the thrill of the speeding train worth the crash that is inevitable? It looks to me like it would go a lot further if you would slow down a little bit, and certainly it would be safer for everybody who lives near the train tracks, which is everybody."

Working with disadvantaged students, on land that was neglected for a long time, has given Strella plenty to contemplate. He has recently turned his energies to the educational farm at Maryland's Pearlstone Center, which focuses on sustainable farming as well as spiritual and social justice work. His conclusions, which he says give him "depthless hope," have been the result of watching both people and landscapes heal.

"In our fields, you will see dandelions, purslane, clover, chicory, amaranth—two or three dozen edible plants that most people think of as weeds," he said. "You can cut through them with knives, but even in spite of that discouragement, four weeks later, they will all be back. I see that as an extremely hopeful thing, that whatever we do next, we haven't exhausted the natural resilience of our soil to heal that land and maintain an abundant system.

"That's also how I see our students," Strella said. "Even in circumstances that are a historical anomaly—with young people growing up and not living close to animals or plants—when the opportunity shows up, their curiosity springs forth. It is not exhausted. It's never exhausted. You can make an argument that their circumstances could have exhausted that, but it hasn't. Those human and ecological reserves that we don't create and can't create— that we in fact get in the way of—are still bigger than us. They are still bigger than our technologies."

Getting Our Hands Dirty

S o here we are, casting our eyes across the American food landscape, and everywhere we look, we see paradoxes. There is a surging interest in small-scale, local food production, and there is a furious consolidation of the biggest food industries in the world. There are powerful popular movements trying to force companies to reveal how they make their food, and highly financed corporate efforts to resist this disclosure. There are gathering efforts to toughen federal safety laws on GMOs and pesticides, and there are outspoken calls—especially during presidential election seasons—to dismantle the EPA altogether.

There are billions of dollars at stake in the United States, and potentially billions of lives at stake in the developing world. And underneath all of these trends is the constant forward march of food technology.

Using a gene-silencing technique called RNA interference, or RNAi, researchers have recently learned to keep apples or potatoes from browning after being harvested—an exciting idea if you are a fast-food company hoping to keep billions of pounds of produce from spoiling on the way to the fryer. Even more dramatic, scientists have invented a new gene-editing technique called CRISPR to edit an organism's genome with ever more impressive precision. DuPont says it will use CRISPR to get drought-tolerant corn and soybeans in fields within five years.

And then there are GM animals. Using cutting-edge gene-editing tools, scientists are learning to create cows without horns, pigs without testicles, and chickens that will produce only female egg-layers. In the Netherlands, researchers have discovered how to turn stem cells from cattle into lab-grown "hamburger."

A company called Aqua Bounty just received approval from the USDA to market a GM salmon, which will grow to marketable size in eighteen months—half the time it would take in nature. Since most farm-raised salmon are fed GM corn and soybeans, we may now, for the first time, have GMOs eating GMOs.

"We're going to see a stream of edited animals coming through because it's so easy," says Bruce Whitelaw, a professor of animal biotechnology at the Roslin Institute at the University of Edinburgh. "It's going to change the societal question from 'If we could do it, would we want it?' to 'Next year we will have it, will we allow it?'"

These advances have excited scientists, industry, and even some animal rights activists. They argue, for example, that creating cows without horns means the animals won't injure each other during confinement and calves won't have to have their horns burned off with a hot iron. Pigs without testicles will produce tastier meat and will mean pigs won't need to be castrated. The grand prize—growing meat from cell cultures rather than from actual living livestock—could mean all kinds of potentially powerful changes to industrial agriculture. We wouldn't need pesticide-laden GM corn, industrial slaughterhouses, or gasoline, because we wouldn't be feeding, slaughtering, or shipping animals around the country. We also wouldn't need to deal with the mountains (or lakes) of animal waste that contaminate our water, or the clouds of methane that contribute to climate change. And we wouldn't need to kill billions of animals to satisfy our bottomless desire for protein.

In 2008, the animal rights group People for the Ethical Treatment of Animals (PETA) offered a $1 million prize to anyone who could make a chicken in a test tube. "If you can grow the chicken flesh from a few cells, that's a lot of birds that won't be suffering," said Ingrid Newkirk, the group's leader, noting that Americans currently eat one million chickens an hour. Ditto for the lab-grown hamburgers. "It is a real burger made of real meat," she said. "It's as real as real can be. The thing that is different about it is that it is not from a filthy slaughterhouse, but from a sterile laboratory."

Outside the kitchen, genetic engineers have been making global headlines breeding everything from mosquitoes to trees. By altering the genes of the *Aedes aegypti* mosquito, a vector for a variety of dangerous diseases, scientists hope to stop the march of the Zika virus, which causes serious neurological damage and birth defects in babies. While the potential impact of GMO mosquitoes on broader ecological systems is still being explored, insects engineered to be sterile, for example, could eliminate the need to return to well-known dangers like aerial spraying South American swamps and jungles with DDT.

Further north, genetic engineers are hoping to restore the American chestnut, once a royal member of the Eastern forest, whose mid-century population—thanks to a fungus from Asia—was reduced by fully 4 billion trees. After nearly twenty-five years of experimenting, biologists may have finally achieved blight resistance by introducing a gene from a wheat plant. What impact GM trees will have on the American landscape remains unknown, but restoring chestnuts to their former range has long been seen as a kind of reforesting gold ring.

As with so many of the promises of genetic engineering, it's hard to argue with some of these possibilities. Changes that could make even incremental dents in our industrial food system—or restore deep and broad damage to our ecosystem—are certainly worth exploring. On the other hand, as Wes Jackson and many others have pointed out, it's worth remembering that most technology is designed to fix problems caused by previous technologies. We have had decades of very exciting agricultural innovation, and we have had decades of dramatically compromised environmental and human health. As inspiring as technological advances may sound, there is always the danger of developing a collective belief that—no matter how badly we screw things up—science will somehow manage to provide a safety net. Such thinking, whether it is conscious or unconscious, can serve to absolve us of taking responsibility for our own ignorance and our own behavior.

To my mind, our learned dependence on (not to say addiction to) industrial food technology has had consequences that are both ecological and philosophical, and include a growing detachment from some of the most fundamental components of life.

Like where a potato comes from.

Nancy Bentley, a friend of mine who owns an organic farm near the Delaware–Maryland line, brings potatoes to public schools to show kids something about what they eat. Very few of the kids have ever actually seen a potato before; to them, a potato is a "chip" that comes in a bag. Needless to say, when they see flecks of dirt on the potato, they recoil in disgust.

If a child thinks a potato is a fried, golden, symmetrical chip in a vacuum-sealed bag, then learns that a potato is actually a dirty brown malignancy that comes from underground, naturally they would prefer the former and reject the latter. And if they can eat the chip without ever having to confront the dirty malignancy, so much the better.

And so it has become for most of us. Our food is now so uniform, so packaged, that if a potato—or a carrot, or a hamburger—looks even slightly different from the last one we tried, we either won't buy it or we will throw it away. In the United States, misshapen vegetables are thrown out in unimaginable numbers by consumers, line cooks, and supermarkets alike. Farmers discard potatoes too small to harvest mechanically. Supermarkets toss cases of hummus or chocolate one day past their expiration date. Kids throw away 40 percent of their school lunches. A recent story in *National Geographic* reported in the United States alone, retailers and consumers throw away 133 billion pounds of food every year. That's "billion" with a *b*.

Such waste may not, in itself, be a "GMO problem," but it is symptomatic of a food system that GMOs have dramatically amplified. In just a few short decades, as we have turned over the growing of our food to a handful of companies, and a handful of crops, we have chosen to be fed according to strictures of efficiency and marketing, not of taste, nutrition, or personal (or environmental) health.

Pesticides and herbicides are not (technically) "GMO problems" either, any more than food waste is, but these compounds are similarly symptomatic of industrial food and scaled up by GMOs. A new study reports that in the United States, the use of glyphosate—sprayed on millions upon millions of agricultural crops—has risen 300-fold since the advent of GMOs; Americans sprayed more than 2.4 billion pounds of the herbicide in the last decade alone. Glyphosate is now sprayed on crops closer and closer to the time of harvest, which means it is showing up in more food and in greater quantities. Rather than crack down on glyphosate, the EPA has approved a doubling of "tolerable" herbicide residues on soybeans, a 49-fold increase on tolerable residues on corn grain, and a 2,000-fold increase in residues on alfalfa grown for animal feed. Glyphosate routinely shows up in everything from soy sauce to breast milk, and new studies are finding links between the compound and cancer, as well as problems with liver, kidney, and metabolic function. Glyphosate has also, it's worth repeating, recently been declared a "probable human carcinogen" by the World Health Organization.

It seems plain that food technology like genetic engineering holds both great promise and great peril. But especially for a culture that has so completely taken its eyes off of farming and food production generally, there is another side to this equation, a side that more directly involves shifting our own role as food consumers.

Perhaps, beyond advances in science and engineering, we need to remember our own traditions of growing and eating food. There was a time—from 10,000 years ago until just a couple of generations ago—when we managed to feed ourselves from a wide variety of plants and animals without the aid of huge companies, and without laying waste to our land or our communities. No matter what else you say about the way we eat today, one thing is clear: we eat differently from how any other people in the history of the world have eaten. Even if we accept the potential benefits of advanced technology—and there are many—we ought also to reintroduce ourselves to the simple act of making a meal.

In my own small way, I've been trying. For the last couple of years, I have required my students to spend time every week working on Fairweather Farm, an organic operation run by my friend Nancy Bentley. My students are mostly humanities majors, not food science majors; mostly future teachers, not future farmers. Admittedly, for my university, the project has been a bit eccentric. I'm not sure I can remember the last time an English professor asked for and received grant money to buy shovels, hoes, and rakes.

My students and I first showed up in Nancy's greenhouses in February 2015. We started by sifting soil and planting broccoli and cabbage and Swiss chard in dozens of plastic trays. March had us preparing beds outside and planting beets and carrots and peas. In April, we harvested asparagus and planted potatoes, and started moving seedlings from the greenhouse to the outside beds. By May, we were transplanting tomatoes and peppers and eggplant and to-matillos and summer squash.

By August and September, when a new crop of students returned to school, we were harvesting crates of vegetables, but we were also pulling up wagonloads of thistle, feeding the sheep, and tending the chickens. In October, we were baling hay, and in November, we were preparing beds for winter.

Week after week, my students wrote journals about their experi-ences. They talked about what it felt like to do manual labor, some of them for the first time in their lives. They talked about what it felt like to get thistle thorns in their hands, or hay in their eyes, or manure on their shoes. But mostly they wrote about the joy they experienced working outside, in the sun or the rain, talking to their friends and their farmer, and getting to know the sheep, and the chickens, and Waldo the goat. They learned about compost, and mushroom soil, and cover crops. They got to see what potatoes look like when they come out of the ground—dirty!—and how good green salsa tastes if you make it from tomatillos you grew yourself.

"The feeling that I got every time I went to the farm is unex-plainable," a student named Danielle wrote. "The satisfaction that I now have when I take a bite out of an organic tomato is something that I would have never experienced if I didn't take this class. This semester changed me as a person. I now think about everything I

do, everything I see, and everything I eat in a completely new perspective."

A student named Hannah wrote that before taking the class she
was "a huge foodie. I loved cooking, finding recipes, and even had
a job at home where I got to do both of these things with relative
freedom. I never once gave thought to where my food came from,
and frankly I never even thought to care about it."

Hannah's sense, vividly confirmed by virtually all of her class-
mates, is that the gap that exists between people and their food is
not just nutritional. It is existential. College students, like the rest
of us, feel profoundly disconnected from some of the most fundamental components of their lives. Ask a roomful of twenty-year-
olds how many of them can tell you precisely how or where their last
meal was grown, and you will get a roomful of blank stares. Ditto if
you ask them where the heat—or the light—in their classroom came
from, or the water in their shower, or the wood in their homes.

Ask them how many generations of human beings have been so
ignorant about these things—food, light, heat, water, and shelter—
and you will begin to have a real conversation. Nancy's farm, for
these students, offers a chance to unplug from their electronic lives,
to feel the warmth of the soil, to go home deeply tired and deeply
renewed.

Wes Jackson calls this work—physical, communitarian, and
ancient—"walking the beans." Tally up all the labor that human
beings have done in all our history—building roads, constructing
cities, fighting wars—and you'll find that we have spent more time
doing one thing than anything else:

Pulling weeds.

Strange, given that most of us never do it anymore. We have no
need, since we don't grow our own food. We live in an era when

physical labor is broadly devalued: we hire other people to do our work *for* us. Indeed, it may seem counterintuitive for Wes Jackson to recommend that we spend more time doing something as (literally) mundane as pulling weeds from vegetable beds. Isn't manual work what technology was invented to replace?

But for Jackson, persuading people to relearn the value of manual work is on a par with replacing 50 million acres of annual wheat with 50 million acres of perennial Kernza: it would be a paradigm shift, a game changer. Growing more of our own food would stitch us back to our land, reintroduce us to our physical bodies—maybe even help repair decades of alienation from the most fundamental things in our lives.

"Nobody likes to walk the beans anymore," Jackson told me. "Instead they use Roundup so—what?—they can go to the gym and jog on a treadmill? Walking the treadmill is okay, hoeing the beans is not? What we need is more eyes per acre."

Getting "more eyes per acre" is precisely the goal of my Literature of the Land class. There is a whole lot more to learning than you can find in a book or in a classroom. Indeed, every small farmer I spoke to during the course of my research—whether they supported GMOs or rejected them—agreed that closing the gap between people and their food is very long overdue, both for our land and for ourselves.

"In our time, our consciousness has to keep evolving," Steffen Schneider, the farmer at Hawthorne Valley, told me. "Unless we learn to work with this inner landscape, I think we will continue to have this huge gap. It's something that's bubbling up everywhere. People working on the land—on the one hand, I know how difficult it is to do every single day, but I think if you work with living nature, there is tremendous inspiration to be gotten from reading and discovering natural phenomena. Once one starts looking at things that way, your work becomes very fruitful. It gives your work a whole different context and purpose.

"That's where we have to start. When I see this amazing enthusiasm to get back to the land—there's this yearning that drives it. If you look at agriculture as a purely economic industry, then one of the primary parameters is efficiency. But this is clearly not complete thinking, because you are dealing with nature, with a living planet, and if you're trying to grow living food, with healthy qualities, it's different than making a shoe or a car. It's a qualitatively different environment. It would be therapeutic for both people and the land if more people did this sort of work."

Certainly this has proved true for my students.

"I, as I imagine many of my peers did, found solace in the farm," a student named Kelsey wrote. "This was where I found peace, where the only material goods I ever required were the occasional shovel or hoe. It was great to go home feeling sore because you had just spent the last two hours pulling weeds, leaving behind a clean bed ready for planting. It was here that I learned so practically the significance of food that you grow yourself. I learned the true meaning of patience and its reward. The joy of eating something picked just moments beforehand, planted perhaps weeks or months beforehand, is indescribable: only understood through experiencing it yourself."

In my class, the idea is not to turn students into farmers, though in recent years a surprising number of them have gone on to work on farms after they graduate. This was not happening five years ago. Now it is. The reason, as far as I can tell, is that students are hungry—literally, hungry—to know more about where their food comes from. They see their generation's relationship with food as emblematic of their engagement with the world generally. Eating microwaved chicken nuggets from the nearest fast-food joint seems considerably less appealing once you have spent dozens of hours tending an organic garden plot—not to mention the friendships you may have developed with a flock of charismatic hens, who trail alongside you as you weed a bed of potatoes.

"Before this semester, I had never worked on a farm, just the occasional community garden or nature preserve," a student named Meghan wrote. "Now, after these past fourteen weeks, it will feel bizarre to go even a week or two *without* spending a few hours in that environment. For the first time, my understanding of environmental issues is not based only on articles I've read from newspapers and vague notions I have about sustainability, but also personal experience and several incredible works of literature.

"On a more interior level, my growth this semester can be harder to see. For me, it happened more slowly. But as I think about it, maybe it shows in the simple fact that feeding cows cabbage at eight-thirty a.m. was the highlight of my entire week, as was kneeling in a greenhouse pulling weeds, talking with Tanya, Rodger, and others about the philosophical things that always seem to come up when your hands are in the dirt."

ACKNOWLEDGMENTS

Constructing a book about a topic as complex and diffuse as our industrial food system has required a great deal of assistance, and I am obliged to a long list of people for their help. During the course of my research, I spoke to dozens of scientists, farmers, activists, and philosophers, from Maryland to Kansas to Hawaii. I also dug deeply into the published work of scores of scientists and journalists who have wrestled with these questions for years, and whose work has shaped my own thinking a great deal.

At the University of Delaware, I have been blessed for twenty years with smart and imaginative students, who have helped me work through a long list of entangled environmental questions. This project owes a special debt to my students in the program in Environmental Humanities, and especially to my research assistants, Kerry Snyder, Tanya Krapf, and Molly Gartland.

Also at Delaware, Blake Meyers opened his plant science laboratory to me and offered far more patience and guidance than any writer deserves. He has since moved on to the Danforth Center in St. Louis, where he joins Jim Carrington, Nigel Taylor, and Paul Anderson, who generously shared their work and expertise with me. Karla Roeber graciously helped organize my visit to the Danforth Center.

In Maryland, thanks to farmers Drew Norman, Joan Norman, Nancy Bentley, Greg Strella, Jennie Schmidt, and Hans Schmidt; and to Sheila Kincaide, Larry Bohlen, and Alfred Sommer, professor emeritus and former dean of the Bloomberg School of Public Health at Johns Hopkins. In Baltimore, my dear friend Arnob Banerjee, MD, PhD, offered his deep expertise on genetics.

In Pennsylvania, thanks to Brian Snyder, head of the Pennsylvania Association for Sustainable Agriculture, and to farmer and entrepreneur Steve Groff.

In Kansas, thanks to Wes Jackson, Tim Crews, Lee DeHaan, and Pheonah Nabukalu. Their work at The Land Institute remains a beacon of environmental integrity.

In Hawaii, thanks to Dennis Gonsalves, Alberto Belmes, Richard Manshardt, Paul Achitoff, Craig Malina, Gary Hooser, Elif Beall, Jeri DiPietro, Fern Rosenstiel, Klayton Kubo, Dustin Barca, Gerry Herbert, Nancy Redfeather, Margaret Wille, Alika Atay, Gerry Ross, Janet Simpson, and Autumn Ness.

In New York, thanks to Steffen Schneider, Conrad Vispo and Claudia Knab-Vispo, and Craig Holdrege, whose work at Hawthorne Valley Farm, the Farmscape Ecology Program, and the Nature Institute, respectively, serves as the very model of enlightened land stewardship.

In Chicago, thanks to Naseem Jamnia, who proved an astute and scrupulous copy editor.

For their help with my understanding of plant genetics and the history of industrial agriculture, I am indebted to a long list of scientists and science writers, whose personal counsel or published work has helped clarify my own. These include especially Evaggelos Vallianatos, John Fagan, John Vandermeer, Marion Nestle, Alfredo Huerta, Bruce Blumberg, David Pimentel, Philip Landrigan, David Mortensen, Steven Druker, Michael Pollan, Nathanael Johnson, Aldo Leopold, Wendell Berry, Carey Gillam, Tom Philpott, Peter Pringle, Pamela Ronald, Richard Manning, Marie-Monique Robin, and Michael Hansen.

At Avery, thanks to Caroline Sutton, Brittney Ross, Brianna Flaherty, and especially Brooke Carey for helping me envision and shape such an unwieldy project. Copy editor Jennifer Eck polished the manuscript's rough edges. And thanks, again and always, to my agent and old friend, Neil Olson.

At home, my deepest gratitude remains reserved for Katherine, Steedman, and Annalisa, who not only accompanied me on research trips to Hawaii and Kansas but also joined me for a 4,000-mile road trip across America's amber waves of grain. Their love, support, and goodwill sustain me every day. They are my life's greatest blessing.

NOTES

Prologue

2 **Tom Brokaw said the tomato:** Michael Winerip, "You Call That a Tomato?" *New York Times,* June 24, 2013, http://www.nytimes.com/2013/06/24/booming/you-call -that-a-tomato.html?smid=tw-share&_r=1.

4 **Genetically modified wheat:** Philip Jones, "While Popularity Eludes GE Foods, AgBiotech Companies Shift Tactics," *Information Systems for Biotechnology* (May 2014), http://www.isb.vt.edu/news/2014/May/Jones.pdf.

5 **In fact, just one-half:** Anne Weir Schechinger, "Feeding the World: Think U.S. Agriculture Will End World Hunger? Think Again," Environmental Working Group, October 5, 2016, http://www.ewg.org/research/feeding-the-world.

6 **This trend has given rise:** "Global Agrochemicals Industry 2014–2019: Trend, Profit and Forecast Analysis," *PR NewsWire* (May 26, 2015), http://www.thestreet.com /story/12911088/1/global-agrochemical-industry-2014-2019-trends-profits-and-fore cast-analysis.html; Marie-Monique Robin, *The World According to Monsanto: Pollution, Corruption and the Control of Our Food Supply* (New York: The New Press, 2010), 5; "Pesticides in Paradise: Hawaii's Health and Environment at Risk," Hawaii Center for Food Safety (May 2015), http://www.centerforfoodsafety.org/files/pesti cidereportfull_86476.pdf.

7 **The World Health Organization recently declared glyphosate:** Lizzie Dearden, "One of World's Most Used Weedkillers 'Possibly' Causes Cancer, World Health Organization Says," *Independent,* June 23, 2015, http://www.independent.co.uk/ news/science/one-of-worlds-most-used-weedkillers-possibly-causes-cancer-world -health-organisation-says-10338363.html.

8 **In a nod to anxious mothers:** Oliver Nieburg, "Hershey's Milk Chocolate and Kisses to Go Non-GM," *Confectionery News,* March 2, 2015, http://www.confectionery news.com/Ingredients/Hershey-in-non-GMO-and-no-high-fructose-corn-syrup -pledge?utm_source=AddThis_twitter&utm_medium=twitter&utm_campaign =SocialMedia#.VPYk_vvu—s.twitter.

8 **Cheerios are made mostly of oats:** "Great-Granddaughter of General Mills Founder Urges Company to Stop Using GMOs," *Friends of the Earth,* Oct. 1, 2014, http:// www.foe.org/news/archives/2014-10-great-granddaughter-of-general-mills-founder -urges-c. General Mills' position did not sit well with Harriet Crosby, an heir to the company fortune, who wrote to its board urging that the company stop using GMOs altogether. GMOs "are only good for big biotech companies like Monsanto that sell both the genetically engineered seeds and the pesticides they are designed to tolerate," Crosby wrote. "The promises of biotechnology are yet unrealized, especially the erroneous claim that they require fewer pesticides. Just the opposite is true. I believe that General Mills can become an even better, more profitable com-

pany by taking global leadership in producing healthy, wholesome, good food without GMOs."

9 **In early 2015, thousands of Polish farmers:** Sophie McAdam, "Anti-GMO Protests Rock Poland as Farmers Demand Food Sovereignty Rights," True Activist, March 4, 2015, http://www.trueactivist.com/anti-gmo-protests-rock-poland-as-farmers-demand -food-sovereignty-rights/.

Chapter 1

19 **These techniques are no more risky:** B. S. Ahloowalia, M. Maluszynski, and K. Nichterlein, "Global Impact of Mutation-Derived Varieties," *Euphytica* 135, no. 2 (April 2014): 187–204; Pamela Ronald and Raoul Adamchak, *Tomorrow's Table: Organic Farming, Genetics, and the Future of Food* (Oxford: Oxford University Press, 2008), 89.

21 **There is truth on both sides of this debate:** Tamar Haspel, "Genetically Modified Foods: What Is and Isn't True," *Washington Post*, October 15, 2013, https://www .washingtonpost.com/lifestyle/food/genetically-modified-foods-what-is-and-isnt -true/2013/10/15/40e4fd58-3132-11e3-8627-c5d7de0a046b_story.html; Tamar Haspel, "The GMO Debate: Five Things to Stop Arguing," *Washington Post*, October 27, 2014, http://www.washingtonpost.com/lifestyle/food/the-gmo-debate-5-things-to-stop -arguing/2014/10/27/e82bbc10-5a3e-11e4-b812-38518ae74c67_story.html. For more on Earth Open Source, see http://earthopensource.org. For more on "GMO Answers," see http://GMOAnswers.com.

22 **"The quest for greater certainty":** Nathanael Johnson, "The Genetically Modified Food Debate: Where Do We Begin?" *Grist*, July 8, 2013, http://grist.org/food/the -genetically-modified-food-debate-where-do-we-begin/.

23 **"no adverse health effects attributed to genetic engineering":** National Research Council and Institute of Medicine of the National Academies, *Safety of Genetically Engineered Foods: Approaches to Assessing Unintended Health Effects* (Washington, DC: National Academies Press, 2004), http://www.nap.edu/openbook.php?record _id=10977&page=8.

23 **"Contrary to popular misconceptions":** American Academy for the Advancement of Science Board of Directors, "AAAS: Labeling of Genetically Modified Foods," October 20, 2012, http://archives.aaas.org/docs/resolutions.php?doc_id=464.

23 **"have passed risk assessments in several countries":** World Health Organization, "WHO Answers Questions on Genetically Modified Food," http://www.who.int/ mediacentre/news/notes/np5/en/.

23 **"There is no more risk in eating GMO food":** Jeremy Fleming, scientific adviser to the European Commission, "No Risk With GMO Food, Says EY Chief," EurActive.com, July 24, 2012, http://www.euractiv.com/section/science-policymaking/news/no-risk -with-gmo-food-says-eu-chief-scientific-advisor/.

24 **A recent meta-analysis of studies:** A. L. Van Eenennaam and A. E. Young, "Prevalence and Impacts of Genetically Engineered Feedstuffs on Livestock Populations," *Journal of Animal Science* 92, no. 10 (May 28, 2014).

26 **"we all know what can happen":** John Vandermeer, "Discovering Science," FoodFirst .org, January 8, 2013, http://www.gmwatch.org/news/archive/2013/14571-professor -john-vandermeer-challenges-lynas-on-gmos.

26 **a short-term (31-day) study:** Maria Walsh et al., "Effects of Short-Term Feeding of Bt MON810 Maize on Growth Performance, Organ Morphology and Function in Pigs," *British Journal of Nutrition* 107 (2012): 364–371, http://journals.cambridge.org/download.php?file=%2FBJN%2FBJN107_03%2FS0007114511003011a.pdf&code=c23ec46ee6bbe8ab3592b187924f0996.

26 **A two-year study of pigs:** Judy Carman et al., "A Long-Term Toxicology Study on Pigs Fed a Combined Genetically Modified (GM) Soy and GM Maize Diet," *Journal of Organic Systems* 8, no. 1 (2013): 38–54, http://www.organic-systems.org/journal/81/8106.pdf; Judy Carman, "Evidence of GMO Harm in Pig Study," GMO Judy Carman, June 5, 2013, http://gmojudycarman.org/new-study-shows-that-animals-are-seriously-harmed-by-gm-feed.

28 **Huerta's skepticism is well founded:** See, for instance, Claire Hope Cummings, *Uncertain Peril: Genetic Engineering and the Future of Seeds* (Boston: Beacon Press, 2008), 41. See also Richard Lacey's testimony in Alliance for Bio-Integrity et al. v. Donna Shalala et al., U.S. District Court, Civil Action No. 98-1300 (CKK), May 28, 1998, http://www.saynotogmos.org/scientists_speak.htm.

29 **residues of the compound routinely show up in British bread:** Arthur Neslen, "EU Scientists in Row over Safety of Glyphosate Weedkiller," *Guardian*, January 13, 2016, http://www.theguardian.com/environment/2016/jan/13/eu-scientists-in-row-over-safety-of-glyphosate-weedkiller.

29 **A study by David Mortensen:** Natasha Gilbert, "Case Studies: A Hard Look at GMO Crops," *Nature* 497, no. 7447 (May 1, 2013): 24–26, http://www.nature.com/news/case-studies-a-hard-look-at-gm-crops-1.12907.

29 **a recent report by the German Federal Institute for Risk Assessment:** "The BfR Has Finalised Its Draft Report for the Re-evaluation of Glyphosate," Bundesinstitut für Risikobewertung, http://www.bfr.bund.de/en/the_bfr_has_finalised_its_draft_report_for_the_re_evaluation_of_glyphosate-188632.html.

30 **President George H. W. Bush appointed:** Robin, *The World According to Monsanto*, 187.

30 **During the Obama administration:** Isabella Kenfield, "Michael Taylor: Monsanto's Man in the Obama Administration," Organic Consumers Association, August 14, 2009, https://www.organicconsumers.org/news/michael-taylor-monsantos-man-obama-administration; Marcia Ishii-Eiteman, "98 Organizations Oppose Obama's Monsanto Man, Islam Siddiqui, for US Agricultural Trade Representative," Organic Consumers Association, February 22, 2010, https://www.organicconsumers.org/news/98-organizations-oppose-obamas-monsanto-man-islam-siddiqui-us-agricultural-trade-representative.

31 **"From the 1940s to the dawn":** E. G. Vallianatos with McKay Jenkins, *Poison Spring: The Secret History of Pollution and the EPA* (New York: Bloomsbury, 2014), ix.

31 **the debate over the safety of farm chemicals:** Dearden, "One of World's Most Used Weedkillers 'Possibly' Causes Cancer."

31 **The net result?:** Wilhelm Klumper and Matin Qaim, "A Meta-Analysis of the Impacts of Genetically Modified Crops," *PLoS ONE* 9, no. 11 (November 2014): e111629, doi:10.1371/journal.pone.0111629; Charles Benbrook, "Impacts of Genetically Engineered Crops on Pesticide Use in the US—The First Sixteen Years," *Environmental Sciences Europe* 24 (2012), doi:10.1186/2190-4715-24-24.

32 "The majority of food": David Pimentel, "Environmental and Economic Costs of the Application of Pesticides Primarily in the United States," *Environment, Development and Sustainability* (2005) 7:229–252.

32 of the six hundred pesticides now in use: Vallianatos with Jenkins, *Poison Spring*, 29.

32 And those numbers tabulate just: David Pimentel et al., "Assessment of Environmental and Economic Impacts of Pesticide Use," in David Pimentel and Hugh Lehman, eds., *The Pesticide Question: Environment, Economics and Ethics* (New York: Chapman & Hall, 1993), 51.

33 Some scientists wonder: Anthony Samsel and Stephani Seneff, "Glyphosate, Pathways to Modern Diseases II: Celiac Sprue and Gluten Intolerance," *Interdisciplinary Toxicology* 6, no. 4 (December 2013): 159–184, http://www.ncbi.nlm.nih.gov/pmc/articles/PMC3945755/; Anthony Samsel and Stephanie Seneff, "Glyphosate, Pathways to Modern Diseases III: Manganese, Neurological Diseases, and Associated Pathologies," *Surgical Neurology International* 6 (March 24, 2015), http://www.ncbi.nlm.nih.gov/pmc/articles/PMC4392553/.

33 This, then, is not a question: Nancy L. Swanson, André Leu, Jon Abrahamson, and Bradley Wallet, "Genetically Engineered Crops, Glyphosate and the Deterioration of Health in the United States of America," *Journal of Organic Systems* 9, no. 2 (2014), http://www.organic-systems.org/journal/92/JOS_Volume-9_Number-2_Nov_2014 -Swanson-et-al.pdf.

34 "It got to the point where some farmers": Quoted in Gilbert, "Case Studies: A Hard Look at GMO Crops," http://www.nature.com/news/case-studies-a-hard-look-at -gm-crops-1.12907.

35 The EPA's recent decision: Philip J. Landrigan and Charles Benbrook, "GMOs, Herbicides, and Public Health," *New England Journal of Medicine* 373 (August 20, 2015): 693–695, doi: 10.1056/NEJMp1505660.

36 Relying on the seed and chemical companies: McKay Jenkins, "Coming Soon: Major GMO Study (Shhh, It Will Be Done in Secret by Russians)," *Huffington Post*, December 18, 2014, http://www.huffingtonpost.com/mckay-jenkins-phd/coming-soon -major-gmo-stu_b_6344812.html.

37 In 1996, the German division: Diahanna Lynch and David Vogel, "The Regulation of GMOs in Europe and the United States: A Case Study of Contemporary European Regulatory Politics," Council on Foreign Relations Press, April 5, 2001, http://www .cfr.org/agricultural-policy/regulation-gmos-europe-united-states-case -study-contemporary-european-regulatory-politics/p8688.

37 "GMOs are dead": Peter Pringle, *Food Inc.: Mendel to Monsanto—The Promises and Perils of the Biotech Harvest* (New York: Simon & Schuster, 2003), 16.

37 More recently, nineteen members: "Majority of EU Nations Seek Opt-Out from Growing GMO Crops," Reuters, October 4, 2015, http://www.reuters.com/article/2015/10/04/eu-gmo-opt-out-idUSL6N0M0IF620151004#qT5acaZpFMvUoIzp.97.

38 "The GMO issue is something": Stephanie Strom, "FDA Takes Issue with the Term 'Non-GMO,'" *New York Times*, November 21, 2015, http://www.nytimes.com/2015/11/21/business/fda-takes-issue-with-the-term-non-gmo.html.

39 Companies have responded aggressively: Eric Lipton, "Food Industry Enlisted Academics in GMO Lobbying War, Emails Show," *New York Times*, September 5, 2015,

http://www.nytimes.com/2015/09/06/us/food-industry-enlisted-academics
-in-gmo-lobbying-war-emails-show.html?_r=0; Jacob Bunge, "Monsanto CEO: 'We
Need to Do More,'" *Wall Street Journal*, January 28, 2014, http://blogs.wsj.com/corporate
-intelligence/2014/01/28/monsanto-ceo-we-need-to-do-more-to-win-gmo-debate/.

39   The story repeated itself: Molly Ball, "Want to Know If Your Food Is Genetically
     Modified?" *Atlantic*, May 14, 2014, http://www.theatlantic.com/politics/archive/2014/
     05/want-to-know-if-your-food-is-genetically-modified/370812/.

40   Almost 90 percent of scientists: Cary Funk and Lee Rainie, "Public and Scientists'
     Views on Science and Society," Pew Research Center, January 29, 2015, http://www
     .pewinternet.org/2015/01/29/public-and-scientists-views-on-science-and-society/.

41   recent Nielsen study . . . "There's no doubt that the industry": Stephanie Strom,
     "Many GMO-Free Labels, Little Clarity over Rules," *New York Times*, January 30,
     2015, http://www.nytimes.com/2015/01/31/business/gmo-labels-for-food-are-in-high
     -demand-but-provide-little-certainty.html.

41   "The sad truth is many": "GMOs and Your Family," Non-GMO Project, http://
     www.nongmoproject.org/learn-more/gmos-and-your-family/.

42   The Non-GMO Project: Strom, "FDA Takes Issue with the Term 'Non-GMO.'"

42   Pro-labeling groups: Ronni Cummins, "'QR' Barcodes: The Latest Plot to Keep
     You in the Dark About GMOs," Organic Consumers Association, October 28, 2015,
     https://www.organicconsumers.org/essays/%E2%80%98qr%E2%80%99-barcodes-
     latest-plot-keep-you-dark-about-gmos; Andrew Kimbrell, "Obama's GMO Embar-
     rassment: Why the New Labeling Bill Just Signed Into Law Is a Sham," *Salon*,
     August 7, 2016, http://www.salon.com/2016/08/07/obamas-gmo-embarrassment-why
     -the-new-labeling-bill-just-signed-into-law-is-a-sham_partner/.

43   As with the regulation: "Big Food's 'DARK Act' Introduced in Congress," Environ-
     mental Working Group, April 9, 2014, http://www.ewg.org/release/big-food-s-dark
     -act-introduced-congress.

43   Opponents of the measure: Ibid.

44   the law accomplished most of what Big Food desired: Kimbrell, "Obama's GMO
     Embarrassment." See also Ramona Bashshur, "FDA and Regulations of GMOs,"
     American Bar Association Health eSource 9, no. 6, February 2013.

Chapter 2

47   In the 1950s alone, some 10 million people: Adam Rome, *The Bulldozer in the Coun-
     tryside* (Cambridge, England: Cambridge University Press, 2001), 123; McKay Jen-
     kins, "The Era of Suburban Sprawl Has to End. So, Now What?" *Urbanite*, May 30,
     2012; "How Long Is the Interstate System?" Federal Highway Administration,
     https://www.fhwa.dot.gov/interstate/faq.cfm#question3.

50   These new foods were cheap: Eric Schlosser, *Fast Food Nation* (New York: Harper
     Perennial, 2002), 3; Katherine Muniz, "20 Ways Americans Are Blowing Their
     Money," *USA Today*, March 24, 2014, http://www.usatoday.com/story/money/
     personalfinance/2014/03/24/20-ways-we-blow-our-money/6826633/.

50   As industrial farms continued to grow: "Report: Number of Animals Killed in US

Increases in 2010," Farm Animal Rights Movement (FARM), http://farmusa.org/statistics11.html.

50 **As farms consolidated and grew:** Cary Fowler and Pat Mooney, *Shattering: Food, Politics and the Loss of Genetic Diversity* (Tucson: University of Arizona Press, 1990), 81.

50 **Over the course of the twentieth century:** USDA National Agricultural Statistics Service, "Field Crops," http://www.nass.usda.gov/Charts_and_Maps/Field_Crops/.

51 **Wallace was right:** Jack Kloppenberg, *First the Seed: The Political Economy of Plant Biotechnology 1492–2000* (Cambridge, England: Cambridge University Press, 1998), 283, cited by Max John Pfeiffer, "The Labor Process and Capitalist Development of Agriculture," *Rural Sociologist* 2, no. 2 (1982): 72–80.

52 **Suddenly, farmers (and their crops):** Michael Pollan, *The Omnivore's Dilemma: A Natural History of Four Meals* (New York: Penguin, 2006), 45.

52 **A similar pattern emerged . . . Today, the six top:** "The World's Top 10 Pesticide Firms—Who Owns Nature?" Organic Consumers Association, November 1, 2008, https://www.organicconsumers.org/news/worlds-top-10-pesticide-firms-who-owns-nature.

53 **This transition, from wartime chemicals:** Pollan, *The Omnivore's Dilemma*, 43; Jill Richardson, "How Monsanto Went from Selling Aspirin to Controlling Our Food Supply," *TruthOut*, April 21, 2013, http://www.truth-out.org/news/item/15856-how-monsanto-went-from-selling-aspirin-to-controlling-our-food-supply.

53 **True, it cranked up:** Balu Bumb and Carlos Baanante, "World Trends in Fertilizer Use and Projections to 2020," International Food Policy Research Institute, 2020 Brief 38, October 1996, http://ageconsearch.umn.edu/bitstream/16353/1/br38.pdf; Smil quoted in Carl Jordan, *An Ecosystem Approach to Sustainable Agriculture* (New York: Springer, 2013), 51.

54 **Similar changes were under way:** Steven Lipin, Scott Kilman, and Susan Warren, "DuPont Agrees to Purchase of Seed Firm for $7.7 Billion," *Wall Street Journal*, March 15, 1999, http://www.wsj.com/articles/SB921268716949898331.

55 **Monsanto rejected the bid:** Jacob Bunge, "Monsanto Rejects Bayer Merger Offer, Says It's Open to Talks," *Wall Street Journal*, May 25, 2016, http://www.wsj.com/articles/monsanto-rejects-bayer-merger-offer-says-its-open-to-talks-1464110057.

55 **"This is an important moment in human history":** Quoted in Peter Pringle, *Food, Inc.: Mendel to Monsanto—The Promises and Perils of the Biotech Harvest* (New York: Simon & Schuster, 2003), 116.

56 **DNA was known to be:** Ania Wieczorek and Mark Wright, "History of Agricultural Biotechnology: How Crop Development has Evolved," *Nature, Education Knowledge* 3, no. 3 (2012): 9–15.

57 **In the late 1980s:** Joe Entine and XioaZhi Lim, "Cheese: The GMO Food Die-Hard GMO Opponents Love (and Oppose a Label For)," GMO Literacy Project, May 15, 2015, http://www.geneticliteracyproject.org/2015/05/15/cheese-gmo-food-die-hard-gmo-opponents-love-and-oppose-a-label-for/.

58 **Monsanto's most important push:** Robin, *The World According to Monsanto*, 138–142.

58 **"It was like the Manhattan Project":** Daniel Charles, *Lords of the Harvest: Biotech, Big Money, and the Future of Food* (Cambridge, MA: Perseus Books, 2002), 67.

58 **It took four years:** Robin, *The World According to Monsanto*, 139.

59 **Americans have become very comfortable:** Alyssa Battistoni, "Americans Spend Less on Food Than Any Other Country," *Mother Jones*, February 1, 2012, http://www.motherjones.com/blue-marble/2012/01/america-food-spending-less.

60 **Because this system has become:** Luke Anderson, *Genetic Engineering, Food and Our Environment* (New York: Chelsea Green, 1999), 70; Stefan Lovgren, "One-Fifth of Human Genes Have Been Patented, Study Reveals," *National Geographic News*, October 13, 2005, http://news.nationalgeographic.com/news/2005/10/1013_051013_gene _patent.html; Matthew Albright, "The End of the Revolution," Council for Responsible Genetics, 2002, http://www.councilforresponsiblegenetics.org/ViewPage.aspx ?pageId=168.

61 **It is also evident:** "U.S. Regulation of Genetically Modified Crops," *Case Studies in Agricultural Biosecurity*, Federation of American Scientists, http://fas.org/biosecurity/ education/dualuse-agriculture/2.-agricultural-biotechnology/us-regulation-of -genetically-engineered-crops.html.

61 **But in reality:** Doug Gurian-Sherman, "Holes in the Biotech Safety Net: FDA Policy Does Not Assure the Safety of Genetically Engineered Foods," Center for Science in the Public Interest, http://www.cspinet.org/new/pdf/fda_report__final.pdf.

62 **U.S. policy "tends to minimize":** Emily Marden, "Risk and Regulation: U.S. Regulatory Policy on Genetically Modified Food and Agriculture," *Boston College Law Review* 44, no. 3 (May 2003), https://www.bc.edu/content/dam/files/schools/law/ lawreviews/journals/bclawr/44_3/02_TXT.htm. Marden gives an excellent summary of FDA, USDA, and EPA regulatory history.

62 **"Our concern for the possible":** Paul Berg, "Potential Biohazards of Recombinant DNA Molecules," *Proceedings of the National Academy of Sciences* 71, no. 7 (July 1974): 2593–2594.

62 **After Berg's letter was published:** Marcia Barinaga, "Asilomar Revisited: Lessons for Today," *Science* 28, no. 5458 (March 2000): 1584.

63 **James Watson, one of the discoverers:** James Watson and John Tooze, *The DNA Story: A Documentary History of Gene Cloning* (San Francisco: W. H. Freeman, 1981), 49.

63 **Watson had nothing but contempt:** Quoted in Diane B. Dutton, *Worse Than the Disease: Pitfalls of Medical Progress* (Cambridge, England: Cambridge University Press, 1992), 195, 327.

64 **"In the 1970s, we were all trying":** Quoted in Steven Druker, *Altered Genes, Twisted Truth* (Salt Lake City: Clear River Press, 2015), 37.

64 **"As genetic engineering became":** Susan Wright, *Molecular Politics: Developing American and British Regulatory Policy for Genetic Engineering, 1972–1982* (Chicago: University of Chicago Press, 1994), 107.

64 **If nothing else:** Schlosser, *Fast Food Nation*, 206.

65 **"The unintended effects cannot":** Quoted in Druker, *Altered Genes, Twisted Truth*, 135.

65 **The director of the FDA's Center for Veterinary Medicine:** Ibid., 133–135.

66 **"This technology is being promoted":** Suzanne Wuerthele, quoted in Jeffrey Smith, "An FDA-Created Health Crisis Circles the Globe," quoted ibid., 186.

67 **"it was clearer than ever that the careers":** Quoted in Druker, *Altered Genes and Twisted Truth*, 132.

67 **Despite this backbeat:** Memorandum from David Kessler to the Secretary for Health and Human Services, March 20, 1992, quoted in Robin, *The World of Monsanto*, 259.

For more on the effectiveness of federal oversight on GMOs, see William Freese and David Schubert, "Safety Testing and Regulation of Genetically Engineered Foods," *Biotechnology and Genetic Engineering Reviews* 21 (November 2004), http://www.centerforfoodsafety.org/files/freese_safetytestingandregulationofgeneticallyebgineeredfoods_nov212004_62269.pdf.

67 **The FDA policy made it official:** "Statement of Policy: Foods Derived from New Plant Varieties," Federal Register 57, no. 104, sec. VI (May 29, 1992): 22991.

67 **Genetic manipulation was no different:** Kessler's comments are from Warren Leary, "Cornucopia of New Foods Is Seen As Policy on Engineering Is Eased," *New York Times*, May 27, 1992, http://www.nytimes.com/1992/05/27/us/cornucopia-of-new-foods-is-seen-as-policy-on-engineering-is-eased.html.

68 **The victory of agribusiness:** Kurt Eichenwald, Gina Kolata, and Melody Petersen "Biotechnology Food: From the Lab to a Debacle," *New York Times*, January 25, 2001, http://www.nytimes.com/2001/01/25/business/25FOOD.html.

68 **Such policy "will speed up":** Quoted ibid.

68 **"What Monsanto wanted (and demanded)":** Druker, *Altered Genes, Twisted Truth*, 138.

69 **In 2002, a committee of the National Academy of Sciences:** National Research Council, "Environmental Effects of Transgenic Plants: The Scope and Adequacy of Regulation," National Academies Press, 2002, http://www.ncbi.nlm.nih.gov/books/NBK207495/. For more on the effectiveness of federal oversight on GMOs, see Freese and Schubert, "Safety Testing and Regulation of Genetically Engineered Foods."

70 **The agency's own policy states:** Freese and Schubert, "Safety Testing and Regulation of Genetically Engineered Foods." And take, for example, the letter the FDA sent to Monsanto about a strain of GMO corn the company hoped to take to market: "Based on the safety and nutritional assessment you have conducted, it is our understanding that Monsanto has concluded that corn products derived from this new variety are not materially different in composition, safety, and other relevant parameters from corn currently on the market, and that the genetically modified corn does not raise issues that would require premarket review or approval by FDA," the FDA's letter said. "As you are aware, it is Monsanto's responsibility to ensure that foods marketed by the firm are safe, wholesome and in compliance with all applicable legal and regulatory requirements."

70 **Companies are not always forthcoming:** Nathanael Johnson, "The GM Safety Dance: What's Rule and What's Real," *Grist*, July 10, 2013, http://grist.org/food/the-gm-safety-dance-whats-rule-and-whats-real/.

70 **For four decades, the American legal system:** E. Freeman, "Seed Police? Part 4," Monsanto.com (November 10, 2008), http://www.monsanto.com/newsviews/Pages/Seed-Police-Part-4.aspx; E. Freeman, "Farmers Reporting Farmers, Part 2," Monsanto.com (October 10, 2008), http://www.monsanto.com/newsviews/Pages/Farmers-Reporting-Farmers-Part-2.aspx; Jessica Lynd, "Gone with the Wind: Why Even Utility Patents Cannot Fence in Self-Replicating Technologies," *American University Law Review* 62, no. 3 (2013): 681–682; "Why Does Monsanto Sue Farmers Who Save Seeds?" Monsanto.com, http://www.monsanto.com/newsviews/pages/why-does-monsanto-sue-farmers-who-save-seeds.aspx.

73 "The truth is Percy Schmeiser": "Percy Schmeiser," a case summary, Monsanto .com, http://www.monsanto.com/newsviews/pages/percy-schmeiser.aspx.

73 This is in direct contrast: "In-Depth: Genetic Modification: Percy Schmeiser's Battle," *CBC News*, May 21, 2004; Phil Bereano and Martin Phillipson, "Goliath vs. Schmeiser: Canadian Court Decision May Leave Multinationals Vulnerable," *GeneWatch* 17, no. 4 (July–August 2004); Roger McEowen and Neil Harl, "Key Supreme Court Ruling on Plant Patents," *Ag Decision Maker Newsletter*, March 2002, https:// www.extension.iastate.edu/agdm/articles/harl/HarlMar02.htm.

74 It's not just that such company-directed: John Vandermeer and Ivette Perfecto, "The AgroEcosystem: A Need for the Conservation Biologist's Lens," *Conservation Biology* 11, no. 3 (June 1997).

75 Without broader research: "Fields of Gold: Research on Transgenic Crops Must Be Done Outside Industry If It Is to Fulfill Its Early Promise," *Nature* 497, no. 7447 (May 1, 2013). "How FDA Regulates Food from Genetically Engineered Plants," U.S. Food and Drug Administration, http://www.fda.gov/Food/FoodScienceResearch/ GEPlants/ucm461831.htm.

Chapter 3

80 Complete sequences (the "chapters"): "DNA, Genes, and Chromosomes," University of Leicester, http://www2.le.ac.uk/departments/genetics/vgec/highereducation/ topics/dnageneschromosomes.

80 All together, an organism's chromosomal chapters: "Learn.Genetics," Genetic Science Learning Center, University of Utah, http://learn.genetics.utah.edu. This website offers a useful interactive graphic showing everything from the size of nucleotides and other cellular material to the mechanics of epigenetics.

81 A five-year project called ENCODE: Francie Diep, "Friction over Function: Scientists Clash on the Meaning of ENCODE's Genetic Data," *Scientific American* (April 2013), http://www.scientificamerican.com/article/friction-over-function-encode/; Claire Robinson, Michael Antoniou, and John Fagan, *GMO Myths and Truths*, 3rd ed. (London: Earth Open Source, 2015): 21–22.

82 The answer lies in the way genes are expressed: Eric Simon, Jean Dickey, and Jane Reece, *Essential Biology* (Boston: Pearson, 2013). I have taken much of the description of gene transcription and translation from this excellent text.

83 Once the transcript of the RNA: Siwaret Arikit et al., "An Atlas of Soybean Small RNAs Identifies Phased siRNAs from Hundreds of Coding Genes," *Plant Cell* 26, no. 12 (December 2014): 4584–4601, http://www.plantcell.org/content/early/2014/12/02 /tpc.114.131847.abstract.

84 Some research suggests that even mental health: Ronald and Adamchak, *Tomorrow's Table*, 159.

84 Exploring these complexities: Adam Thomas, "Maize Genomics," University of Delaware's *UDaily*, March 2, 2015, http://www.udel.edu/udaily/2015/mar/maize -reproduction-030215.html.

92 Fagan is a molecular biologist: Brandon Copple, "Scientist, Activist, Yogi?" *Forbes*, October 30, 2000, http://www.forbes.com/forbes/2000/1030/6612054b.html.

97 **How you feel about GMOs:** Joel Achenbach, "Why Do So Many Reasonable People Doubt Science?" *National Geographic*, March 2015, http://ngm.nationalgeographic.com/2015/03/science-doubters/achenbach-text.

99 **The study, by Gilles-Éric Séralini:** Gilles-Éric Séralini et al.,"Long Term Toxicity of a Roundup Herbicide and a Roundup-Tolerant Genetically Modified Maize," *Food and Chemical Toxicology* 50, no. 11 (November 2012): 4221–4231, http://www.sciencedirect.com/science/article/pii/S0278691512005637 (where it is now labeled "Retracted"); Druker, *Altered Genes, Twisted Truths*, 302.

99 **Jean-Marc Ayrault, France's prime minister:** Andrew Pollack, "Paper Tying Rat Cancer to Herbicide Is Retracted," *New York Times*, November 28, 2013, http://www.nytimes.com/2013/11/29/health/paper-tying-rat-cancer-to-herbicide-is-retracted.html; "Smelling a Rat," *Economist*, December 7, 2013.

100 **Almost instantly, the journal was deluged:** Jon Entine, "Séralini Threatens Lawsuit in Wake of Retraction of Infamous GMO Cancer Rat Study," *Forbes*, November 29, 2013, http://www.forbes.com/sites/jonentine/2013/11/29/notorious-seralini-gmo-cancer-rat-study-retracted-ugly-legal-battle-looms/; Kate Kelland, "Journal Withdraws Controversial French Monsanto GMO Study," Reuters, November 29, 2013, http://www.reuters.com/article/science-gm-retraction-idUSL2N0JE0FM20131129; "GMO Study Retracted: Censorship of Caution?" *Living on Earth*, December 6, 2013, http://loe.org/shows/segments.html?programID=13-P13-00049&segmentID=2. See also letters to the editor posted at http://www.sciencedirect.com/science/article/pii/S0278691512005637.

100 **A more in-depth look:** "Elsevier Announces Article Retraction from Journal *Food and Chemical Toxicology*," Elsevier.com, November 28, 2013, http://www.elsevier.com/about/press-releases/research-and-journals/elsevier-announces-article-retraction-from-journal-food-and-chemical-toxicology#sthash.KgeQj4lq.dpuf.

100 **More than a hundred scientists . . . "The retraction is erasing":** "Scientists Pledge to Boycott Elsevier," *Ecologist*, December 5, 2013, http://www.theecologist.org/blogs_and_comments/commentators/2187010/scientists_pledge_to_boycott_elsevier.html.

101 **The decision was based "not on the grounds":** Brian John, "Letter to the Editor," *Food and Chemical Toxicology* 65 (March 2014): 391, http://www.sciencedirect.com/science/article/pii/S0278691514000040.

102 **Goodman's "fast-tracked appointment":** Claire Robinson and Jonathan Latham, "The Goodman Affair: Monsanto Targets the Heart of Science," *Independent Science News*, May 20, 2013, https://www.independentsciencenews.org/science-media/the-goodman-affair-monsanto-targets-the-heart-of-science/.

102 **Brian John offered a sharp answer:** John, "Letter to the Editor," *Food and Chemical Toxicology*.

102 **"When those with a vested interest":** "Seralini [*sic*] and Science: An Open Letter," *Independent Science News*, October 2, 2012, http://www.independentsciencenews.org/health/seralini-and-science-nk603-rat-study-roundup/.

103 **The issues raised by the Séralini study:** Jenkins, "Coming Soon: Major GMO Study (Shhh, It Will Be Done in Secret by Russians)."

104 **Blumberg's own work:** "Chemicals That Promote Obesity down the Generations," *Living on Earth*, January 18, 2013, http://loe.org/shows/segments.html?programID=13-P13-00003&segmentID=1.

104 **Looking over one of Blumberg's:** Author interview with Bruce Blumberg.

Chapter 4

110 **By the early 1990s, ringspot:** Dennis Gonsalves, "Transgenic Papaya in Hawaii and Beyond," *AgBioForum* 7, nos. 1 & 2 (2004): 36–40, http://www.agbioforum.org/v7n12/v7n12a07-gonsalves.pdf.

112 **The mid-1980s was an exciting time:** Harold Schmeck, "Plants 'Vaccinated' Against Virus," *New York Times*, May 6, 1986, http://www.nytimes.com/1986/05/06/science/plants-vaccinated-against-virus.html. Eventually, such work on virus resistance would lead to the discovery of "RNA silencing," one of biology's major advances in the past two decades. Silencing RNA has been used to develop a treatment for macular degeneration and is considered a promising field for science leading to therapies for both plants and animals. The technique won a Nobel Prize for Americans Andrew Fire and Craig Mello in 2006.

113 **Following Beachy's lead:** Ronald and Adamchak, *Tomorrow's Table*, 159. For more on how silencing RNAs work.

113 **"It is rather rare that a potential solution":** Gonsalves, "Transgenic Papaya in Hawaii and Beyond," 37.

115 **One year later:** Jennifer Mo, "The Man Behind the Rainbow," *Biofortified*, June 21, 2012, http://www.biofortified.org/2012/06/rainbow/.

116 **"a tireless innovator" . . . "is a model":** Quoted in Paul Voosen, "Crop Savior Blazes Biotech Trail, but Few Scientists or Companies Are Willing to Follow," *New York Times*, September 21, 2011, http://www.nytimes.com/gwire/2011/09/21/21greenwire-crop-savior-blazes-biotech-trail-but-few-scien-88379.html?pagewanted=all.

118 **As evidence of GMO contamination:** Steven Layne, "Thailand's GMO Experiment, Part 2," *Phuket News*, June 10, 2014, http://www.thephuketnews.com/thailand-gmo-experiment-part-2-46788.php.

119 **There have also been problems:** Melanie Bondera, "Papaya and Coffee: GMO 'Solutions' Spell Market Disaster," in Hawaii SEED, *Facing Hawai'i's Future: Essential Information About GMOs*, 2nd ed. (Koloa, Hawaii: Hawaii SEED, 2012), 48–50.

121 **There are signs that Gonsalves's message:** Voosen, "Crop Savior Blazes Biotech Trail."

Chapter 5

126 **Because the fields themselves:** Pesticide Action Network, "Television Show and Body Testing Confirm Children's Exposure to Neurotoxic Pesticide," February 4, 2016, http://www.panna.org/press-release/television-show-and-body-testing-confirm-children%E2%80%99s-exposure-neurotoxic-pesticide; Anita Hofschneider, "Syngenta Workers Seek Medical Aid After Pesticide Use on Kauai," *Honolulu Civil Beat*, January 22, 2016, http://www.civilbeat.com/2016/01/syngenta-workers-seek-medical-aid-after-pesticide-use-on-kauai/.

127 **Land use on Kauai:** Hank Soboleski, "Pablo Manlapit and the Hanapepe Massacre," *Garden Island*, September 10, 2006, http://thegardenisland.com/news/pablo-manlapit-and-the-hanapepe-massacre/article_57bc7ca1-a576-5c2f-8ad9-1a7641eb4c21.html.

128 **The "experiments" taking place:** "Pesticides in Paradise," Hawaii Center for Food Safety.

131 **Kauai's sole pesticide inspector:** Paul Koberstein, "GMO Companies Are Dousing

Hawaiian Island with Toxic Pesticides," *Grist*, June 16, 2014, http://grist.org/business-technology/gmo-companies-are-dousing-hawaiian-island-with-toxic-pesticides/; Sophie Cocke, "Frustrated by State's Inactivity, Kauai County Takes Pesticide Fight Into Its Own Hands," *Huffington Post*, October 8, 2013, http://www.huffingtonpost.com/2013/10/08/kauai-county-gmo-fight_n_4064787.html.

131 **The disclosure records "are believed to contain":** Letter from Thomas Matsuda to Gary Hooser, July 22, 2014.

132 **A Hawaii Department of Agriculture (HDOA) log:** State of Hawaii Department of Agriculture Inspection Log, Kauai, 2011–2012.

133 **State records show:** Restricted Use Pesticides Sold on Kauai, 2010–2012, Hawaii Department of Agriculture.

133 **Other records show:** "Pesticide Use by Large Agribusiness on Kauai: Findings and Recommendations of the Joint Fact Finding Study Group," March 2016, http://www.accord3.com/docs/GM-Pesticides/draft-report/JFF%20Full%20Report%20-%20DRAFT.pdf.

133 **A study published in March 2014:** Philippe Grandjean and Philip Landrigan, "Neurobehavioural Effects of Developmental Toxicity," *Lancet* 13, no. 3 (March 2014): 330–338, http://www.thelancet.com/journals/laneur/article/PIIS1474-4422(13)70278-3/abstract.

134 **Recent hair sample testing:** Pesticide Action Network, "Television Show and Body Testing Confirm Children's Exposure to Neurotoxic Pesticide."

134 **But it wasn't just chlorpyrifos:** Ibid.

134 **Also in the cocktail: permethrin . . . chlorpyrifos:** See "The Food We Eat: An International Comparison of Pesticide Regulations," David Suzuki Foundation, October 2006, http://www.davidsuzuki.org/publications/downloads/2006/DSF-HEHC-Food1.pdf.

134 **"even more acutely toxic":** E. G. Vallianatos with McKay Jenkins, *Poison Spring: The Secret History of Pollution and the EPA* (New York: Bloomsbury, 2014), 39.

134 **Another ingredient in the cocktail:** Rachel Aviv, "A Valuable Reputation," *New Yorker*, February 10, 2014, http://www.newyorker.com/magazine/2014/02/10/a-valuable-reputation.

135 **"had their heads in their hands":** Amanda Gregg, "'Stink Weed' Sends Some Home from Waimean School," *Garden Island*, November 15, 2006, http://thegardenisland.com/news/stink-weed-sends-some-home-from-waimea-school/article_5a66ba3d-2194-53dc-8bda-807e794625ae.html.

135 **Though it is eaten:** Ibid.; Adam Harju, "Odor Investigation Ongoing," *Garden Island*, November 18, 2006, http://thegardenisland.com/news/odor-investigation-ongoing/article_650db9f3-06a5-5960-8f47-97a433eac349.html; Amanda Gregg, "Report Reveals Discrepancies in Spraying Incident," *Garden Island*, March 12, 2007, http://thegardenisland.com/news/report-reveals-discrepancies-in-spraying-incident/article_046c3b40-4ecb-54f6-a1b9-e90d3c032153.html.

136 **Company claims about stinkweed:** "Pesticides in Paradise," Hawaii Center for Food Safety.

136 **There was no questioning . . . There was no evidence:** Memo from J. Milton Clark to Peter Adler, chairman of Task Force on Kauai Pesticides and GMO, May 19, 2015,

included in "Pesticides Use by Large Agribusinesses on Kauai." See http://www
.accord3.com/docs/GM-Pesticides/report/JFF%20Report%20Errata.pdf.

137 **What the island needed:** Paul Achitoff, "GMOs in Kauai: Not Just Another Day in
Paradise," *Huffington Post*, March 5, 2014, http://www.huffingtonpost.com/paul-achitoff/
gmos-in-kauai-not-just-an_b_4899491.html; "AAP Makes Recommendations to Reduce Children's Exposure to Pesticides," American Academy of Pediatrics, November
26, 2012, https://www.aap.org/en-us/about-the-aap/aap-press-room/pages/AAP-Makes
-Recommendations-to-Reduce-Children's-Exposure-to-Pesticides.aspx.

137 **"many qualitative examples"** . . . **"We all share a deep concern":** "Doctors and Nurses
Implore Mayor: Sign Bill 2491 into Law Now!" Stop Poisoning Paradise, http://www
.stoppoisoningparadise.org/#!doctors-and-nurses-letters-to-mayor/csim. Sample letters
from doctors and nurses to Kauai County Council, delivered October 20, 2013.

138 **The doctors' worries reflected:** "August 5, 2013 Kauai County Council Dr Evslin
Kauai Pediatrician," YouTube video, 7:11, posted by "Mom's Hui Kaua'i," August 10,
2013, https://www.youtube.com/watch?v=8g6D8xAz6fA.

138 **Margie Maupin, a nurse practitioner:** "Margie Maupin Nurse Practitioner in Support
Bill 2491," YouTube video, 6:09, posted by "Occupy Hawaii," November 16, 2013,
https://www.youtube.com/watch?v=FrqHU8Y-QCo.

139 **In 2011, more than a hundred of Klayton Kubo's neighbors filed a lawsuit:** For information on the trial and its issues, see Vanessa Van Voorhis, "Waimea Residents Sue
Pioneer," *Garden Island*, December 13, 2011, http://thegardenisland.com/mobile/
article_82ff2c3e-2632-11e1-9ca7-001871e3ce6c.html; Tom LaVenture, "Waimea Residents Suing Pioneer Hi-Bred," *Garden Island*, June 14, 2012, http://thegardenisland
.com/news/local/waimea-residents-suing-pioneer-hi-bred/article_607adf66-b5ff-11e1
-a19b-0014bcf887a.html; Sophie Cocke, "Does Hawaii's Failure to Enforce Pesticide
Use Justify Action by Kauai?" *Honolulu Civil Beat*, October 8, 2013, http://www
.civilbeat.com/2013/10/20066-does-hawaiis-failure-to-enforce-pesticide-use-justify
-kauais-action/; Associated Press, "Jury Awards Kauai Residents over $500K in Dust
Lawsuit," *Maui News*, May 10, 2015, http://www.mauinews.com/page/content.detail/
id/597884/Jury-awards-Kauai-residents-over—500K-in-dust-lawsuit.html?nav=5031.
A year later, residents filed a second suit, claiming Pioneer consistently failed to control
the erosion and pesticide-laden dust from its GMO test fields.

140 **Jervis reminded the court:** Koberstein, "GMO Companies Are Dousing Hawaiian
Island with Toxic Pesticides." A 1,500-page report from the state Department of
Agriculture concluded that although Syngenta did apply pesticides near the school,
it did so correctly, and ruled out the chemical Hi-Tech as the culprit.

140 **"Kaua'i produces more GMO seeds than anyplace":** Vanessa Van Voorhis, "Large-
Scale Die-off of Sea Urchins Discovered off Kaumakani," *Garden Island*, February
23, 2012, http://thegardenisland.com/news/local/large-scale-die-off-of-sea-urchins
-discovered-off-kaumakani/article_16081484-5a1b-11e1-bca7-0019bb2963f4.html.

141 **That same winter . . . Vandana Shiva:** Jon Letman, "Opposition Crops Up to GMO
Foods in Hawaii," Al Jazeera, February 16, 2013, http://www.aljazeera.com/indepth/
features/2013/02/20132514512529904.html.

143 **Industry executives claimed:** Cocke, "Frustrated by State's Inactivity, Kauai County
Takes Pesticide Fight into Its Own Hands."

144 **Companies dismissed complaints:** Letman, "Opposition Crops Up to GMO Foods in Hawaii."

144 **During the hearings on the bill:** Kristine Uyeno, "GMO Public Hearing on Kauai Draws Hundreds," KHON-TV, July 31, 2013, http://khon2.com/2013/07/31/gmo -public-hearing-on-kauai-draws-hundreds/.

145 **Yet within weeks:** Carey Gillam, "Anti-GMO Crop, Pesticide Ballot Initiative Launched in Hawaii," Reuters, February 24, 2014, http://www.reuters.com/article/ 2014/02/25/usa-gmos-hawaii-idUSL1N0LU0A220140225.

145 **An attorney representing:** Nestor Garcia, "Federal Judge Declares New Kauai GMO, Pesticide Law Invalid," KHON-TV, August 25, 2014, http://khon2.com/ 2014/08/25/federal-judge-declares-new-kauai-gmo-pesticide-law-invalid/.

146 **"The big question":** Keoki Kerr, "State, Kauai Set Up Panel to Study GMO Pesticide Impacts," *Hawaii News Now*, December 3, 2014, http://www.hawaiinewsnow.com/ story/27532992/state-kauai-set-up-panel-to-study-gmo-pesticide-impacts.

147 **For local residents:** Associated Press, "Jury Awards Kauai Residents over $500K in Dust Lawsuit."

147 **Ten days after the verdict:** Anita Hofschneider, "DuPont Pioneer Shuts Down One Kauai Facility," *Honolulu Civil Beat*, May 20, 2015, http://www.civilbeat.com/2015/05/ dupont-pioneer-shuts-down-one-kauai-facility/.

147 **"Syngenta did not want me there":** Gary Hooser, "From Kauai to Switzerland— Why We Went, What We Accomplished and What's Next," *GaryHooser's Blog*, https://garyhooser.wordpress.com/2015/05/03/from-kauai-to-switzerland-why-we -went-what-we-accomplished-and-whats-next/.

Chapter 6

149 **The Big Island, basically, had one:** Nathanael Johnson, "Here's Why Hawaii's Anti-GMO Laws Matter," *Grist*, November 20, 2014, http://grist.org/food/heres -why-hawaiis-anti-gmo-laws-matter/.

155 **Big agricultural companies:** Anita Hofschneider, "Hawaii Farmers, Biotech Industry Challenge Big Island's GMO Ban," *Honolulu Civil Beat*, June 9, 2014, http://www .civilbeat.com/2014/06/hawaii-farmers-biotech-industry-challenge-big-islands -gmo-ban/.

156 **Industry representatives were elated:** Associated Press, "Federal Judge Rules Against Big Island GMO Law," *NewsOK*, November 26, 2104, http://newsok.com/federal -judge-rules-against-big-island-gmo-law/article/feed/765175.

158 **Despite unparalleled weather . . . Hawaii's agricultural experts:** "Pesticides in Paradise," Hawaii Center for Food Safety.

162 **For the benefit of present and future generations:** The Constitution of the State of Hawaii, Article XI, http://lrbhawaii.org/con/conart11.html.

163 **It was a safe bet:** "Hawaiians Take on Monsanto and GMOs," Pachamama Alliance, June 25, 2014, http://www.pachamama.org/webcasts/hawaiians-take-on-monsanto -gmos.

164 **The petition also urged voters:** "A Bill Placing a Moratorium on the Cultivation of Genetically Modified Organisms," Chapter 20.39 of the Maui County Code, http://www.mauicounty.gov/Archive/ViewFile/Item/19197.

165 **The ball was now:** Wendy Osher, "Maui Petition Filed Against GMO Industry,

Monsanto Responds," *Maui Now*, April 8, 2014, http://mauinow.com/2014/04/08/maui-petition-filed-against-gmo-industry-monsanto-responds/.

166 **A Monsanto employee . . . Another Monsanto employee:** "Employees Rally in Support of Monsanto on Maui and Molokai—4/3/14," YouTube video, 3:13, posted by "Maui Now," https://www.youtube.com/watch?v=VOU7K7Fr5Zw.

167 **Rather than try to convince:** "Maui County Genetically Modified Organism Moratorium Initiative (November 2014): Arguments Against," Ballotpedia, http://ballotpedia.org/Maui_County_Genetically_Modified_Organism_Moratorium_Initiative_(November_2014) #Arguments_against; "Over 11,000 Maui County Citizens Stand Up and Say 'Nuff Already' to Biotech Experimentation with a History Making Social Action," PRWeb, April 14, 2014, http://www.prweb.com/releases/2014/04/prweb11758961.htm.

168 **The companies also flexed:** For the original language of the initiative and for the text of the final moratorium question, respectively, see "Maui County Genetically Modified Organism Moratorium Initiative (November 2014): Full Text," Ballotpedia, https://ballotpedia.org/Maui_County_Genetically_Modified_Organism_Moratorium_Initiative_(November_2014),_full_text; "Maui County Genetically Modified Organism Moratorium Initiative (November 2014): Ballot Question," Ballotpedia, http://ballotpedia.org/Maui_County_Genetically_Modified_Organism_Moratorium_Initiative_(November_2014)#Ballot_question.

171 **Industry advertisements—typically attributed:** "Maui County Genetically Modified Organism Moratorium Initiative (November 2014): TV Ads," Ballotpedia, http://ballotpedia.org/Maui_County_Genetically_Modified_Organism_Moratorium_Initiative_(November_2014)#TV_ads.

171 **The campaign finance reports:** For campaign spending reports, see "Maui County Genetically Modified Organism Moratorium Initiative (November 2014): Campaign Finance," Ballotpedia, http://ballotpedia.org/Maui_County_Genetically_Modified_Organism_Moratorium_Initiative_(November_2014), #Campaign_finance.

173 **Lorrin Pang, a Maui physician:** "Over 11,000 Maui County Citizens Stand Up and Say 'Nuff Already' to Biotech Experimentation with a History Making Social Action," PRWeb.

173 **Alika and the SHAKA Movement held:** Anita Hofschneider, "1,000 Votes: Maui GMO Farming Ban Squeaks By," *Honolulu Civil Beat*, November 4, 2014, http://www.civilbeat.com/2014/11/1000-votes-maui-gmo-farming-ban-squeaks-by/.

174 **Sure enough, when the final vote . . . just over 51 percent:** "Maui County Genetically Modified Organism Moratorium Initiative (November 2014): Election Results," Ballotpedia, http://ballotpedia.org/Maui_County_Genetically_Modified_Organism_Moratorium_Initiative_(November_2014)#Election_results.

175 **The next day, Monsanto:** Audrey McAvoy, "Monsanto, Dow Unit Sue Maui County over GMO Law," Associated Press, November 13, 2014, http://www.ksl.com/?nid=1200&sid=32342920.

175 **Kurren reassigned the case:** Audrey McAvoy, "Maui Group Wins Ability to Intervene in GMO Case," Associated Press, December 15, 2014, http://www.stltoday.com/business/local/maui-group-wins-ability-to-intervene-in-gmo-case/article_99140705-b5df-55e9-9cf4-25a4ebc32f86.html.

175 **No portion of this ruling:** "Maui County Genetically Modified Organism Moratorium Initiative (November 2014): Aftermath," Ballotpedia, http://ballotpedia.org/

Maui_County_Genetically_Modified_Organism_Moratorium_Initiative
_(November_2014)##Aftermath.

Chapter 7

186 **In other words, the Danforth Center:** "Board of Directors," Donald Danforth Plant
Science Center, https://www.danforthcenter.org/about/leadership/board-of-directors.

186 **A cynic might claim:** Doreen Stabinsky, "Hearts of Darkness: The Biotech Industry's Exploration of Southern Africa," *GeneWatch* 15, no. 6 (November–December,
2002).

187 **"a stalking horse for corporate proponents":** Phil Bereano, "Bill's Excellent African
Adventure: A Tale of Technocratic Agroindustrial Philanthrocapitalism," *GeneWatch*
26, no. 1 (January–February 2013): 16.

187 **"If companies are going to continue to claim":** Marion Nestle, *Safe Food: Bacteria, Biotechnology, and Bioterrorism* (Berkley: University of California Press, 2003),
p. 247.

188 **"In the United States, we've seen":** Peter Rosset, Frances Moore Lappé, and Joseph
Collins, "Lessons from the Green Revolution: Do We Need New Technology to End
Hunger?" *Tikkun* 15, no. 2 (March/April 2000): 52–56.

191 **Potrykus visualized peasant farmers:** J. Madeleine Nash, "This Rice Could Save a
Million Kids a Year," *Time*, July 31, 2000, http://content.time.com/time/magazine/
article/0,9171,997586,00.html.

192 **In the end, Potrykus and his team:** Pringle, *Food Inc*, 31–35; Amy Harmon, "Golden Rice:
Lifesaver?" *New York Times*, August 24, 2013, http://www.nytimes.com/2013/08/25/
sunday-review/golden-rice-lifesaver.html?_r=0.

192 **The journal *Science* announced:** Mary Lou Guerinot, "The Green Revolution Strikes
Gold," *Science* 287, no. 5451 (January 14, 2000): 241–243.

192 **Greenpeace, which had taken:** "Field of Dreams: Potrykus' Golden Rice," *Financial
Times*, February 25, 2000; see http://www.genepeace.ch/new/2000/fields_of_dreams
_2002.htm.

193 **In an article in *The New York Times Magazine*:** Michael Pollan, "The Great Yellow
Hype," *New York Times Magazine*, March 4, 2011, http://www.nytimes.com/
2001/03/04/magazine/04WWLN.html.

193 **"the 'selling' of vitamin A":** Vandana Shiva, "Golden Rice: Myth, Not Miracle,"
GMWatch, January 12, 2014, http://www.gmwatch.org/news/archive/2014/15250
-golden-rice-myth-not-miracle.

193 **And so it has gone:** Christopher J. M. Whitty, Monty Jones, Alan Tollervey, and
Tim Wheeler, "Biotechnology: Africa and Asia Need a Rational Debate on GM
Crops," *Nature* 497 (May 2, 2013): 31–33.

194 **In August 2013, hundreds of protesters:** Harmon, "Golden Rice: Lifesaver?"

194 **Such research mirrored work:** Frances Moore Lappé, Joseph Collins, and Peter Rosset, with Luis Esparza, *World Hunger: Twelve Myths, Food First* (New York: Grove
Press, 1998), 61.

195 **Borlaug has never been shy:** *Harvest of Fear*, PBS *Frontline/NOVA*, April 23, 2001,
http://www.pbs.org/wgbh/harvest/.

195 **M. S. Swaminathan, a renowned:** M. S. Swaminathan, "Perspective: The Challenges Ahead," *New Agriculturist* 11 (April 1999), http://www.new-agri.co.uk/99-4/perspect.html.

196 **Indeed, for every plant scientist:** David Pimentel, "Changing Genes to Feed the World: A Review of *Mendel in the Kitchen: A Scientist's View of Genetically Modified Foods*, by Nina Federoff and Nancy Marie Brown," *Science* 306, no. 5697 (October 29, 2004): 815.

197 **Catherine Ives, the scientist:** *Harvest of Fear*, PBS *Frontline/NOVA*.

197 **Monsanto boasts that it has already trained:** Jonathan Gilbert, "In Paraguay, the Spread of Soy Strikes Fear in Hearts of Rural Farmers," *Time*, August 9, 2013, http://world.time.com/2013/08/09/in-paraguay-rural-farmers-fear-the-spread-of-soy/; Christine MacDonald, "Green Going Gone: The Tragic Deforestation of the Chaco," *Rolling Stone*, July 28, 2014, http://www.rollingstone.com/culture/news/green-going-gone-the-tragic-deforestation-of-the-chaco-20140728.

198 **Two years later, in 2012:** Simon Romero, "Vast Tracts in Paraguay Forest Being Replaced by Ranches," *New York Times*, March 24, 2012, http://www.nytimes.com/2012/03/25/world/americas/paraguays-chaco-forest-being-cleared-by-ranchers.html; Tracy Barnett, "Paraguay Takes Hard Line on GMOs," *Huffington Post*, September 1, 2010, http://www.huffingtonpost.com/tracy-l-barnett/paraguay-takes-hard-line-_b_701182.html.

198 **In Argentina, meanwhile, a woman:** "Sofia Gatica," The Goldman Environmental Prize, 2012, http://www.goldmanprize.org/recipient/sofia-gatica/.

200 **"The problem is that there are very few":** Alfred Sommer, "Vitamin A Deficiency Disorders: Origins of the Problem and Approaches to Its Control," *AgBioWorld*, 2011, http://www.agbioworld.org/biotech-info/topics/goldenrice/vit_a.html.

201 **If anything, the prospect of climate change:** Felix Chung, "The Search for the Rice of the Future," from a special issue devoted to rice in "Nature Outlook," a supplement to *Nature* 514, no. 7524 (October 30, 2014); Tim Folger, "The Next Green Revolution," *National Geographic*, October 2014, http://www.nationalgeographic.com/foodfeatures/green-revolution/.

201 **In 2015 alone, scientists published:** Leigh Dayton, "Blue Sky Rice," from a special issue devoted to rice in "Nature Outlook," a supplement to *Nature* 514, no. 7524 (October 30, 2014).

203 **In Kenya, farmers have:** Rachel Cernansky, "The Rise of Africa's Super Vegetables," *Nature*, June 9, 2015, http://www.nature.com/news/the-rise-of-africa-s-super-vegetables-1.17712.

203 **Yet in June 2016, more than one hundred Nobel laureates:** Joel Achenbach, "107 Nobel Laureates Sign Letter Blasting Greenpeace over GMOs," *Washington Post*, June 29, 2016, https://www.washingtonpost.com/news/speaking-of-science/wp/2016/06/29/more-than-100-nobel-laureates-take-on-greenpeace-over-gmo-stance/.

Chapter 8

207 **Monsanto began testing Roundup Ready wheat:** Justin Gillis, "Monsanto Pulls Plan to Commercialize Gene-Altered Wheat," *Washington Post*, May 11, 2004,

http://www.washingtonpost.com/wp-dyn/articles/A15998-2004May10.html; Steven Mufson, "Monsanto Shares Fall as South Korea Joins Pause in Wheat Imports," *Washington Post*, May 31, 2013, https://www.washingtonpost.com/business/economy/monsanto-shares-fall-as-south-korea-joins-pause-in-wheat-imports/2013/05/31/5df79a3a-ca2c-11e2-8da7-d274bc611a47_story.html.

208 **Just ask Larry Bohlen:** Marc Kaufman, "Biotech Critics Cite Unapproved Corn in Taco Shells," *Washington Post*, September 18, 2000, https://www.washingtonpost.com/archive/politics/2000/09/18/biotech-critics-cite-unapproved-corn-in-taco-shells/e7973551-d518-47dc-9bdf-d7931e5e8b49/.

209 **The results, in one case:** Andrew Pollack, "Kraft Recalls Taco Shells with Bioengineered Corn," *New York Times*, September 23, 2000, http://www.nytimes.com/2000/09/23/business/kraft-recalls-taco-shells-with-bioengineered-corn.html.

209 **To Larry Bohlen, it was entirely obvious:** *Harvest of Fear*, PBS *Frontline/NOVA*.

210 **Things have only gotten rockier:** Francie Grace, "Anheuser-Busch Starts Rice War," *CBS News*, April 13, 2005, http://www.cbsnews.com/news/anheuser-busch-starts-rice-war/.

211 **Suddenly, it was StarLink all over again:** Richard Vanderford, "Bayer Settles Rice Contamination Suits for $750M," Law360, July 1, 2011, http://www.law360.com/articles/255594/bayer-settles-rice-contamination-suits-for-750m; Steven Mufson, "Unapproved Genetically Modified Wheat from Monsanto Found in Oregon," *Washington Post*, May 30, 2013, https://www.washingtonpost.com/business/economy/unapproved-genetically-modified-wheat-from-monsanto-found-in-oregon-field/2013/05/30/93fe7abe-c95e-11e2-8da7-d274bc611a47_story.html; Jonathan Randles, "Monsanto's Legal Risk Sprouts as Asia Shuns Modified Wheat," Law360, May 31, 2013, http://www.law360.com/articles/446381/monsanto-s-legal-risk-sprouts-as-asia-shuns-modified-wheat; "USDA Announces Close and Findings of Investigation in the Detection of Genetically Engineered Wheat in Oregon in 2013," September 26, 2014, https://www.aphis.usda.gov/newsroom/2014/09/pdf/ge_wheat.pdf.

213 **"There are too many of us":** Wes Jackson, "Commencement Address: The Serious Challenge of Our Time," University of Kansas, May 19, 2013, https://landinstitute.org/wp-content/uploads/2014/04/WJackson-KU-Commencement-Addr_May2013.pdf.

216 **Jared Diamond, the scientist and bestselling author:** Jared Diamond, "The Worst Mistake in the History of the Human Race," *Discover*, May 1987, http://discovermagazine.com/1987/may/02-the-worst-mistake-in-the-history-of-the-human-race.

216 **African Bushmen eat some seventy-five different wild plants:** Ibid.

216 **The Irish were so dependent:** Richard Manning, *Against the Grain* (New York: North Point Press, 2005), 72–79.

219 **Instead of the ecological desert:** Lee DeHaan and David Van Tassel, "Useful Insights from Evolutionary Biology for Developing Perennial Grain Crops," *American Journal of Botany* 101, no. 10 (2014): 1801–1819.

220 **The idea seemed so obvious:** "Biomimicry: Nature's Alternative to Genetically Engineered Foods," *Environment and Ecology* (2015), http://environment-ecology.com/biomimicry-bioneers/372-biomimicry-natures-alternative-to-genetically-engineered-foods-.html.

220 **Deep roots would also mean:** Robert Kunzig, "Perennial Solution," *National Geographic*, April 2011, http://ngm.nationalgeographic.com/2011/04/big-idea/perennial-grains-text.

221 **The work has been slow:** Richard Harris, "Prairie Pioneer Seeks to Reinvent the Way We Farm," National Public Radio, October 21, 2009.

221 **In the early 1900s:** Lance Gibson and Garren Benson, "Origin, History, and Uses of Oat (*Avena sativa*) and Wheat (*Triticum aestivum*)," Iowa State University, Department of Agronomy (rev. January 2002), http://agron-www.agron.iastate.edu/Courses/agron212/Readings/Oat_wheat_history.htm.

222 **Claims from industrial corn companies:** James Conca, "It's Final: Corn Ethanol Is of No Use," *Forbes*, April 20, 2014, http://www.forbes.com/sites/jamesconca/2014/04/20/its-final-corn-ethanol-is-of-no-use/.

Chapter 9

233 **the company has to feed most of the 569 million chickens:** "Look What the Chicken Industry Is Doing for Delmarva," Delmarva Poultry Industry Inc., 2014, https://www.dpichicken.org/faq_facts/docs/FACTS14.pdf.

234 **Given this level of scrupulous attention:** Jennie Schmidt, "The Truth About GMOs," *Boston Review*, September 6, 2013, http://www.bostonreview.net/forum/truth-about-gmos/farmer-choose-gmos.

237 **The Plenish beans have clearly been:** Jennie Schmidt, "GMO Versus Non-GMO: The Cost of Production," *The Foodie Farmer*, December 29, 2014, http://thefoodiefarmer.blogspot.com/2014/12/gmo-versus-nongmo-cost-of-production.html.

237 **"We've had folks ask us":** Marc Gunther, "GMO 2.0: Genetically Modified Foods with Added Health Benefits," *Guardian*, June 10, 2014, http://www.theguardian.com/sustainable-business/2014/jun/10/genetically-modified-foods-health-benefits-soybean-potatoes.

237 **Clearly, the companies that both:** Sandy Bauers, "DuPont Develops New Cooking Oil from Genetically Modified Soybeans," *Philadelphia Inquirer*, January 4, 2015, http://www.philly.com/philly/columnists/sandy_bauers/20150104_GreenSpace__DuPont_develops_new_cooking_oil_from_genetically-modified_soybeans.html.

237 **In the fall of 2014, DuPont Pioneer and Perdue AgriBusiness:** "DuPont Pioneer, Perdue Announce Doubling of Acreage for 2014 Plenish High Oleic Soybean Program," Pioneer press release, November 18, 2013, https://www.pioneer.com/home/site/about/news-media/news-releases/template.CONTENT/guid.17F567F5-5AF2-7ED8-A7D6-27BAA176DEA2.

238 **"We're always looking for ways":** "DuPont Pioneer, Perdue AgriBusiness to Double Acreage for 2015 Plenish High Oleic Soybean Program," Perdue press release, October 24, 2014, http://www.perduefarms.com/News_Room/Press_Releases/details.asp?id=1129&title=DuPont%20Pioneer,%20Perdue%20AgriBusiness%20to%20double%20acreage%20for%202015%20Plenish%AE%20high%20oleic%20soybean%20program; Sean Cloughery, "Delaware Officials Get Behind Popular Plenish Beans," AmericanFarm.com, http://www.americanfarm.com/publications/the-delmarva-farmer/events/1705-delaware-officials-get-behind-popularity-for-plenish-beans.

239 **"There is no 'one' system"**: Jennie Schmidt, "Farming Techniques Do Not Belong to One Farming System," *The Foodie Farmer*, June 5, 2015, http://thefoodiefarmer .blogspot.com/2015/06/farming-techniques-do-not-belong-to-one.html.

239 **"Because this is spraying and not dousing"**: Jennie Schmidt, "Spraying Isn't Dousing," *The Foodie Farmer*, June 15, 2015, http://thefoodiefarmer.blogspot.com/2015/06/ spraying-isnt-dousing.html.

245 **In the Chesapeake Bay watershed**: Timothy Wheeler, "Sodden Fields Delay Planting of Cover Crops to Aid the Bay," *Baltimore Sun*, November 21, 2009, http://articles .baltimoresun.com/2009-11-21/news/0911200158_1_crop-program-planting-busy -harvesting.

248 **Indeed, they note, the word is featured prominently**: "Our Commitment to Sustainable Agriculture," Monsanto.com, http://www.monsanto.com/whoweare/pages/our -commitment-to-sustainable-agriculture.aspx.

248 **"Industrial agriculture today"**: André Leu and Ronnie Cummins, "From 'Sustainable' to 'Regenerative'—The Future of Food," *Common Dreams*, November 10, 2015, http:// www.commondreams.org/views/2015/10/28/sustainable-regenerative-future-food.

Chapter 10

251 **"There are two spiritual dangers"**: Aldo Leopold, *A Sand County Almanac and Sketches Here and There* (1949; Oxford and New York: Oxford University Press, 1989), 6.

254 **But given that by the end of the century**: Whitty, Jones, Tollervey, and Wheeler, "Biotechnology: Africa and Asia Need a Rational Debate on GM Crops."

262 **"Over the last few years"**: "Del. Farmers' Market Sales Double in 5 Years," WBOC News, December 26, 2014, http://www.wboc.com/story/27709471/del-farmers-market -sales-double-in-5-years.

263 **Thanks to a national surge**: Stephanie Strom, "USDA to Start Program to Support Local and Organic Farming," *New York Times*, September 28, 2014, http://www .nytimes.com/2014/09/29/business/usda-to-start-program-to-support-local-and -organic-farming.html.

263 **As helpful as this has been**: Brad Plumer, "The $956 Billion Farm Bill in One Graph," *Washington Post*, January 28, 2014, http://www.washingtonpost.com/news/wonkblog/ wp/2014/01/28/the-950-billion-farm-bill-in-one-chart/.

264 **Gaps can be reduced further**: Sarah Yong, "Can Organic Crops Compete with Industrial Agriculture?" *Berkeley News*, December 9, 2014, http://news.berkeley.edu/2014/ 12/09/organic-conventional-farming-yield-gap/.

264 **A thirty-year study**: Bill Liebhardt, "Get the Facts Straight: Organic Agriculture Yields Are Good," *Organic Farming Research Foundation* 10 (Summer): 1, 4–5; Pimentel, "Changing Genes to Feed the World," 815.

268 **But into this vacuum**: "Mission 2014: Feeding the World," Massachusetts Institute of Technology, http://12.000.scripts.mit.edu/mission2014/solutions/urban-agriculture; Trish Popovich, "10 American Cities Lead the Way with Urban Agricultural Ordinances," *Seedstock*, May 27, 2014, http://seedstock.com/2014/05/27/10-american -cities-lead-the-way-with-urban-agriculture-ordinances/.

Epilogue

275 **Using a gene-silencing technique:** Tom Philpott, "The Seven Biggest Food Stories of 2015," *Mother Jones*, December 30, 2015, http://www.motherjones.com/tom-phil pott/2015/12/here-are-biggest-food-and-farm-stories-2015.

276 **And then there are GM animals:** Maggie Fox, "Lab-Grown Meat Is Here—But Will Vegetarians Eat It?" *NBC News*, August 5, 2013, http://www.nbcnews.com/ health/diet-fitness/lab-grown-meat-here-will-vegetarians-eat-it-f6C10830536; Kat McGowan, "This Scientist Might End Animal Cruelty—Unless GMO Hardliners Stop Him," *Mother Jones*, September–October 2015, http://www.motherjones.com/ environment/2015/07/fahrenkrug-genetic-modification-gmo-animals.

276 **"We're going to see a stream":** Amy Harmon, "Open Season Is Seen in Gene Editing of Animals," *New York Times*, November 26, 2015, http://www.nytimes.com/2015/11/27/ us/2015-11-27-us-animal-gene-editing.html.

279 **A recent story in *National Geographic*:** Elizabeth Royte, "How Ugly Fruits and Vegetables Can Help Solve World Hunger," *National Geographic*, March 2016, http:// www.nationalgeographic.com/magazine/2016/03/global-food-waste-statistics/.

279 **Pesticides and herbicides are not (technically):** See Charles Benbrook, "Trends in Glyphosate Herbicide Use in the United States and Globally," *Environmental Sciences Europe* 28, no. 3 (February 2016), http://enveurope.springeropen.com/articles/10.1186/s12302-016 -0070-0; Mary Ellen Kustin, "Monsanto's Glyphosate Weed-Killer Is Pervasive, GMO Labels Nonexistent," Environmental Working Group, April 10, 2015, http://www.ewg .org/agmag/2015/04/gmo-weed-killer-pervasive-gmo-labels-nonexistent; Mary Ellen Kustin, "Americans at Greater Risk of Glyphosate Exposure Than Europeans," Environmental Working Group, February 3, 2016, http://www.ewg.org/agmag/2016/02/ americans-greater-risk-glyphosate-exposure-europeans.

INDEX

ABOUT THE AUTHOR

McKay Jenkins has been writing about people and the natural world for thirty years. His most recent book, *ContamiNation* (Avery), chronicled his investigation into the myriad synthetic chemicals we encounter in our daily lives, and the growing body of evidence about the harm these chemicals does to our bodies and the environment. His book *Poison Spring* (Bloomsbury, 2014), cowritten with E. G. Vallianatos, has been called "a jaw-dropping exposé of the catastrophic collusion between the Environmental Protection Agency and the chemical industry" (*Booklist*, starred review).

Jenkins's other books include *Bloody Falls of the Coppermine: Madness and Murder in the Arctic Barren Lands* (Random House, 2005); *The Last Ridge: The Epic Story of the U.S. Army's 10th Mountain Division and the Assault on Hitler's Europe* (Random House, 2003); and *The White Death: Tragedy and Heroism in an Avalanche Zone* (Random House, 2000). Jenkins is also the editor of *The Peter Matthiessen Reader* (Vintage, 2000).

A former staff writer for *The Atlanta Constitution*, Jenkins has also written regularly on environmental matters for *The Huffington Post*, *Outside*, *Orion*, *The New Republic*, and many other publications.

He holds degrees from Amherst, Columbia's Graduate School of Journalism, and Princeton, where he received a PhD in English. Jenkins is currently the Cornelius Tilghman Professor of English, Journalism, and Environmental Humanities at the University of Delaware, where he has won the Excellence in Teaching Award. He lives in Baltimore with his family.

ALSO BY McKAY JENKINS

"A *Silent Spring* for the human body."
—RICHARD PRESTON, author of *The Hot Zone*

ContamiNation
My Quest to Survive in
a Toxic World